DOWN FROM THE TREES

*Man's Amazing Transition from
Tree-Dwelling Ape Ancestors*

DOWN FROM THE TREES

*Man's Amazing Transition from
Tree-Dwelling Ape Ancestors*

Ralph D. Hermansen

Apple Academic Press Inc.
3333 Mistwell Crescent
Oakville, ON L6L 0A2
Canada

Apple Academic Press Inc.
9 Spinnaker Way
Waretown, NJ 08758
USA

© 2018 by Apple Academic Press, Inc.

First issued in paperback 2021

No claim to original U.S. Government works

ISBN 13: 978-1-77463-418-9 (pbk)
ISBN 13: 978-1-77188-733-5 (hbk)

Library and Archives Canada Cataloguing in Publication

Hermansen, Ralph D., author
Down from the trees : man's amazing transition from
tree-dwelling ape ancestors / Ralph D. Hermansen, PhD.

Includes bibliographical references and index.
Issued in print and electronic formats.
ISBN 978-1-77188-733-5 (hardcover).--ISBN 978-0-429-46595-6 (PDF)

1. Human evolution. I. Title.

| GN281.H47 2018 | 599.93'8 | C2018-902740-1 | C2018-902741-X |

CIP data on file with US Library of Congress

Apple Academic Press also publishes its books in a variety of electronic formats. Some content that appears in print may not be available in electronic format. For information about Apple Academic Press products, visit our website at **www.appleacademicpress.com** and the CRC Press website at **www.crcpress.com**

CONTENTS

ABOUT THE AUTHOR

 Ralph D. Hermansen, a graduate chemist by education, had a 40-year career in materials science and engineering. He moved around a bit in his early career in order to broaden his experience and knowledge. Ralph liked multidisciplinary projects that were at the leading edge of technology. Consequently, he has worked on the design of an atom smasher, on classified weapons programs, on design of the Boeing 737 and 747 airplanes, on military aircraft and their electronic platforms, on space satellites, on missiles, on automotive electronics, on telephony, and other commercial applications. He planted his roots with Hughes Aircraft Corporation in his later career and eventually retired from there as a Senior Scientist. Concurrent to his daytime job at HAC, Ralph was an instructor for Hughes' off-hour classes, teaching Pascal programming, and materials engineering.

Career-wise, he has contributed numerous technical articles to journals and magazines, has 21 patents, and has published two books related to his career field. The most recent is *Polymeric Thermosetting Compounds:* available from Apple Academic Press. This unique book is about the art and science of custom formulating adhesives, encapsulants, castings, coatings, etc., towards a specified set of property goals.

Ralph adopted a second career path after retiring in 1998, that of multidisciplinary science writer. His primary focus has been evolutionary science, which is incredibly diverse. Anthropology was his favorite subject at the University of New Mexico in the 1960s, but he stayed with materials science/engineering as a surer path to feeding his growing young family. However, New Mexico and the surrounding area are rich in prehistoric sites. Ralph loved to visit Frijoles Canyon, Chaco Canyon,

Aztec, and Mesa Verde to inspect the cliff dwellings of the long gone Anasazi. Years before writing *Down from the Trees*, Ralph had been filling his bookshelves with books on evolutionary topics. He used these sources to research and compose articles on selected aspects of evolution. He exchanged these thoughts with the late author Elaine Morgan through correspondence for years. Human evolution has become controversial when it really need not be. Ralph's goal in writing *Down from the Trees* was to help curious readers learn the facts in plain language without battling the jargon, realize that the evidence is overwhelming, see that our prehistoric past is fascinating and filled with unsolved puzzles, and realize that they can join in the fun of interpreting the new clues being constantly discovered.

LIST OF ABBREVIATIONS

AAH	aquatic ape hypothesis
K	Cretaceous period
PCR	polymerase chain reaction
SLI	specific language impairment
SNP	nucleotide polymorphism
T	tertiary period

ACKNOWLEDGMENTS

This book was made possible due to the help of many people. First is my wife of 40 years, Janet, who is very supportive of my writing. Mark McGuire solves my puzzling computer and software problems for me, but also is a friend who helps critique my writing. Longtime friend Laurie Bluestein and new friends Peggy Allum and Bill Stacy all critiqued the manuscript and offered helpful suggestions.

INTRODUCTION

THE UNDERLYING PROBLEM

We humans are a curious lot. We are the only animals in the world who are capable of wondering about our place in the grand scheme of things. Why are we here? What is our purpose? Why do we have to eventually die? What happens to our loved ones when they die? Where did we come from?

In this book, we are very concerned with that last question, where did we come from? Scientists tell us that we descended from the African apes. In fact, the DNA of humans is 98.8% identical to the DNA of a chimpanzee. Jared Diamond, a Pulitzer Prize-winning author and scientist, suggests that we are actually the third species of chimpanzee. He is referring to the fact that two species of chimpanzees currently exist—the common chimpanzee (*Pan troglodytes*) and the bonobo (*Pan paniscus*). Man would then be the third species of chimpanzee as indicated by the DNA evidence.

I think most people would have trouble in distinguishing between common chimps and bonobos. Both of them are tree-dwelling apes with their bodies adapted for swinging through the trees. For example, their arms are much longer than their legs, their fingers are curved to better grasp branches, and their feet can grasp branches too. They are also different from humans in that they are covered with fur and move like quadrupeds when on the ground. Neither of them can talk, write a sentence, or add up a column of numbers.

If we are the third species of chimps, then we are very different from our ape cousins indeed! We cannot grasp a tree or limb with our feet. Our bodies are adapted to upright walking. All five of our toes are inline. Our legs are longer than our arms, and our hands are precision instruments for making and using tools. We are virtually hairless and have a more efficient method of cooling our brains and bodies than apes do. We have a mechanism of perspiration unknown in the ape world. In many ways, we are physically different from our ape cousins. However, the even bigger difference between humans and apes is in our greater mental capability. Our brains are three times larger than those of apes, and we can do things with our brains that apes cannot even imagine doing. We can speak, write, and have an innovative spirit. We have passed our knowledge down through the generations and built on it continuously. We have used that intelligence to reconstruct the world to our own benefit. All these differences make us want to defy any association with our ape cousins.

So here is the problem: How is it possible that DNA evidence clearly shows that humans belong to the ape family, yet we are clearly quite different from all the apes? The common chimp and bonobo have evolved separately in isolation from each other for a million years; yet, they are very similar to each other. Our common ancestor with the chimps is estimated to have lived about seven million years ago, but it may have been as little as five million years ago. We don't think that the chimp has evolved very much during that time span. Consequently, our ancestors beginning with the common ancestor 5–7 million years ago and ending with the human species today must have undergone almost all the evolution. How is that possible? Answering that question is the goal of this book.

HARD SCIENCE VERSUS SOFT SCIENCE

Science has been applied to many different types of fields—chemistry, physics, biology, astronomy, geology, and psychology are several research areas that come to mind. The so-called hard sciences employ laboratory testing to verify a hypothesis. Another scientist working in another laboratory should be able to repeat the experiment in question and obtain the same results. Our confidence in the experiment's results increases as more and more verifications are made.

Physics and chemistry are the two examples of hard sciences. Some of the other sciences are softer in the sense that laboratory testing is not possible and so other means of investigation must be employed. Take astronomy for example. We cannot approach the Sun and conduct temperature measurements, or visit another galaxy, or go back in time and observe the origin of the Solar System. What we can do is make observations, postulate mechanisms for what we observe, make predictions, and judge the value of our theories on the accuracy of our predictions. For example, predicting the time and date of the next solar eclipse is a test of our planetary motion beliefs expressed in mathematics. As we gain faith in our basic theories, we can use the principles learned to analyze other phenomena.

The field of astronomy has made immense progress using such techniques. Black holes, predicted by Einstein's equations, were once regarded as too fantastic to contemplate. Now, Stephen Hawkings predicts their properties from his wheelchair and writes popular books about them. Astronomers believe that massive black holes lie at the center of most, if not all, galaxies. The activity of stars around the black hole in the center of our Milky Way galaxy is stimulating the curiosity of scientists worldwide. Billions of dollars are being spent to better visualize the first stars born after the Big Bang.

Paleoanthropology is the scientific field that addresses the study of human origins. It is considered to be a soft science because it is usually not possible to test our hypotheses in the laboratory or to go back in time and observe how humans originated and evolved over time. Paleoanthropology is also a multidisciplinary science. We utilize geology, radiochemistry, biology, physiology, zoology, botany, paleontology, and many other sciences to solve ancient riddles. Genetics, which involves the analysis of the DNA code, has recently proven to be a powerful tool. Like the wonders revealed by astronomers, you will be impressed by how much we have deduced about our origins and evolutionary change.

Paleoanthropology is also a combative science. This book is built upon the hard work and discoveries of numerous scientists, but they don't always agree with each other's interpretations. One area of conflict comes from dating ancient fossils and relics. Advancements in dating techniques are helping to resolve these disagreements. Another area of conflict comes from disagreement on species classification and ancestral relationships. New species are continually being discovered, which adds both clarity and confusion to the controversial subject. What was once thought to be a family tree now appears more like a family bush. I do not believe that any of these conflicts are important to the overall story of human evolution. It is the nature of this field that new discoveries and better techniques will change our knowledge base. It is also true that new discoveries provide additional verification to many older conclusions.

THE THREEFOLD APPROACH

This book is the result of my extensive reading on the subject. The bibliography given at the end of this book lists some of the books that I have read in order to construct my case. One of my writing goals is to present the material at a level where an intelligent reader could understand the material even though he/she is unfamiliar with the topic. I have avoided too much detailing or dwelling on a particular topic too long. I was also aware that some people are skeptical of the scientific story of man's origins. So, a secondary goal was to prove what is claimed or at least explain the processes by which scientists arrive at their conclusions. The result was a threefold approach to unravel the past. The first approach examines the fossil record, which includes a list of bipedal walkers of various types that link our common ape ancestors to humans of today. The second approach is the application of the evolutionary theory to interpret the evidence. Evolutionary change is a fact, but many people do not understand how it operates. Evolution can only modify the creature that exists at that time, and it must offer an advantage to the creature at that time. Evolution has no long-term intents. The third approach is the use of DNA science to unveil the prehistoric past. This last approach is newer than the other two approaches and is itself in a state of rapid development. It is also a powerful tool. The discovery of DNA and amazing advances in its testing and analysis have enabled to establish the genetic distance between any two species. In other words, we can calculate how long ago those two species evolved from a common ancestor. Moreover, mutations serve as markers in an individual's DNA. By comparing the markers in the DNA of different individuals, a chronology of the past events can sometimes be constructed. The migration of *Homo sapiens* from Africa to the far corners of the Earth has been tracked using such techniques.

ORGANIZATION OF THE BOOK

This book is organized according to the three main approaches, discussed above.

- Part I provides a background to fossils and their interpretation. It covers the K/T extinction, which ended the reign of the dinosaurs and allowed mammals and birds to repopulate the vanquished Earth. The K/T extinction is the name of the worldwide mass extinction, which occurred at the end of the Cretaceous period (K) and the beginning of the Tertiary period (T). Part I also surveys the existing apes of the world and describes their cultures.

- Part II addresses hominin fossils, which fill the time span between our common ancestor with chimps and humans of today. Hominins are essentially upright standing, bipedal primates. We shall see how they become less apelike and more humanlike over the passage of time. The Australopiths came first. They were essentially apes, which became proficient bipedal walkers, but remained small-brained and sought refuge in the trees. The Homo lineage came next. They scavenged and hunted for meat, became long-legged runners, fashioned themselves and used tools, and became ever more humanlike over time.

- Part III explains the evolutionary theory. Beginning with the contributions of Charles Darwin, this section explains evolution due to natural selection and how genetic science filled the knowledge gaps in the theory and thus became the Modern Synthesis. One chapter discusses Richard Dawkin's contribution to the recent evolutionary theory where genes are considered to be of prime importance. The Red Queen effect is the subject of another chapter. It involves sexual selection and co-evolution. These topics are tied to human evolution in the last chapter of the section.

- Part IV is a basic tutorial on DNA discovery and principles. It covers the race to discover the structure of DNA and explains how that discovery unlocked the secret of life. It answers how DNA exerts control over our bodies by producing specialized proteins. It clarifies the code of life, whereby a four-letter code unlocks messages contained in our genes. It also explains mutations, which function as markers useful in unlocking our unknown prehistory.

- Part V examines a series of problems in human evolution, which were resolved using DNA techniques. How old is our species? Which of the living apes are we most closely related to? What are the migration paths when *Homo sapiens* migrated from Africa and settled in the rest of the world? How closely related are our species and Neanderthals? Did we interbreed with them?

- Part VI addresses the individual differences between man and ape. This list includes bipedal walking, hairlessness, big brains, speech and language, fire, cooking, toolmaking, and finally, sex. This section employs all three tools (fossils, evolutionary principles, and DNA evidence) to interpret the data.

- Part VII addresses more recent human development, including: the Great Leap Forward beginning 50,000 years ago and the discovery of agriculture 10,000 years ago and its progression to civilization through innovation. In the last chapter, the chronology of human evolution from an ape origin to what we are today is reexamined in light of what we learned throughout the book.

INTENDED USE OF THE BOOK

This book was written with the intent of surveying the science underlying our current knowledge of human evolution. It is not intended to be a complete text for each of the many subtopics covered. The chapters have been kept intentionally short to prevent the reader from becoming bored and losing interest. This book would work well as a supplement to the textbooks of anthropology courses. The students can learn about the long evolutionary path leading from tree-dwelling apes to modern humans, with relatively easy reading compared to the typical textbooks. That evolutionary journey began when our ancestral apes descended from the trees. The book will be of interest to curious lay readers as well. The scientific story of our origins is one which should be of interest to all of us.

PART I

FOSSILS TELL A STORY

Our goal in this book is to understand the transformations that we have undergone if we truly descended from tree-dwelling apes. In this section, we shall see how the story of life on the Earth is written in the strata as fossils of once-living plants and animals. Geologists have painstakingly constructed the sequence of events over millions of years. About 1.5 billion years ago, multicelled animals evolved to the point where they left clear and distinct fossils to document their existence. Since that time, a progression of new animals has appeared in the fossil record: first fishes dominated the seas, then amphibians evolved to occupy the land, then reptiles evolved to extend that range to drier regions, and dinosaurs improved on the reptile skeletal design to rule the Earth for an immensely long time.

The so-called K/T extinction event ended the dinosaur reign about 65 million years ago and paved the way for our species' eventual appearance. Mammals and birds replaced the dinosaurs as dominant animals, and new species of each evolved to fill the empty ecological niches. Amongst the tree dwellers, prosimians proliferated at first, but monkeys and apes eventually replaced them.

Today, there are only a handful of ape types left on the Earth, whereas there once were many. We can learn a lot about ourselves by learning more about these living apes. Each of the ape types is examined here. This includes the gibbon, the orangutan, the gorilla, the chimpanzee, and its close relative, the bonobo.

CHAPTER 1

STRATIGRAPHY

CONTENTS

1.1 SCOPE

The story of life on the Earth is there for us to read. It is written in the rocks in the form of fossils. Indeed, scientists have been analyzing such rocks for more than a century, and now, we have a pretty good idea of when and how life advanced from microorganisms into plants and animals. We know because they left images of themselves, in the form of fossils. They are the preserved forms of plant and animal remains, sometimes many millions of years old. They progressively record the first fishes, the first amphibians, the first reptiles, and the first mammals. We are particularly interested in the mammals because we humans are mammals and this book is about our evolutionary story. This chapter introduces the concept of rocky strata as a chronological record of life evolution. It also discusses fossils, what they are, and how they form.

1.2 SEASHELLS ON A MOUNTAINTOP

In the early sixties, I relocated from the green, softly rolling hills of the Illinois dairy farmlands, where I had grown up, to the semi-arid, rugged, and steep Rocky Mountains region of New Mexico. The sunny city of Albuquerque, where I resettled, lies in a huge valley in which the Rio Grande snakes across it at its lowest point. This lowest point is nearly a mile above sea level. The 10,000-foot-high Sandia Mountains lie to the east and seem to tower over the city. The subtler Manzano Mountains lie visibly to the southeast. Mount Taylor, at 80 miles distance, peeks up above the foothills to the west, and finally, you can see the tall mountain range behind Sante Fe to the north on most days of the year. These mountains are close to 100 miles away.

Sandia Crest is the highest peak near Albuquerque, towering above the city at 10,678 feet elevation. It can be accessed by auto, cable car, or hiking trails. The view from the top is astounding. You can easily see for a hundred miles, but you can also find something extraordinary by keeping your eyes glued to the ground under your feet. You can actually find marine fossils, such as seashells, on the top of a tall mountaintop. How did they get to the top of a mountain? The answer is that long ago this mountaintop was an ocean floor. It has been uplifted in the last 10 million years as part of the formation of the Rio Grande Rift. Today, this stratum is two miles higher than where it was when these animals existed. This experience of actually seeing fossil seashells on a mountaintop was a real eye-opener for me. Just imagine seeing evidence of ancient sea life two miles above sea level in the middle of a vast desert area!

1.3 FOSSILS

Have you ever been to the Petrified Forest? Fallen trees, some of them two feet in diameter, lie askew on the Arizona desert floor. Some of them have been sawed through, and the cross-sections exhibit colorful tree rings. However, there is one thing really unusual about these tree trunks; they are actually solid rock. The reason that they are now fossilized rock is their great age. These trees were alive over 200 million years ago in the Triassic period of the Mesozoic era (*Note*: The Mesozoic era consists of three periods: The Triassic, the Jurassic, and the Cretaceous). These trees lived when the first dinosaurs roamed on the land. Over that vast geological time period, minerals have gradually replaced the decomposing wood and created a rock replica of the trees. Thus, we have fossil evidence of a forest that existed very long ago.

Animals can leave fossils in a similar manner. I have a beautiful rock fossil of an ammonite from the Canadian Rockies, which has been cross-sectioned to reveal its chambers. Like the nautilus, the ammonite was a spiral, sea-dwelling animal capable of rising or sinking in water by flooding or emptying its chambers. Unlike the nautilus, which still exists, the ammonite has been extinct for 65 million years. It went extinct at the same time as the dinosaurs (Figure 1.1).

Whereas species of sea animals often leave numerous fossils, land animals do it far less frequently. Dead land animals are usually devoured by a series of scavengers. Decay and weathering remove what is left after the scavengers. Conditions have to be favorable for fossils to form and that is a rare event. Yet it is through fossils that we humans have been able to discover the chronology of countless species of animals that have existed on this Earth. For the purposes of this book, I am interested in one

FIGURE 1.1 An ammonite fossil.

particular animal and just how it came into existence. That animal is us, human beings and our ancestors, and as you will see, there is a rich fossil trail to examine.

1.4   GEOLOGY BECOMES A SCIENCE

Stratigraphy is one of the more important concepts in the field of geology. It is the study of strata or layers of rock or soils. These layers were built-up by deposition. Limestone, for example, is the deposition of sea creature skeletons accumulating over thousands of years. Rock that originates from this sedimentary process is called sedimentary rock. When you see layers of rock having different colors as in the Grand Canyon, the layers (a.k.a. strata) were formed at different times by different-colored minerals depositing and solidifying into rock. Scientists can identify particular strata by the fossils of plants and animals that existed in that period of time. This interests us because these strata can tell us the story of life on Earth. Incidentally, not all strata contain fossils. Optically visible fossils first appeared a little over one-half billion years ago. So strata older than that is without visible fossils.

"Superposition" is the principle that a lower stratum was deposited earlier than a higher stratum and is therefore older than the higher stratum. In other words, the lower you go in a series of strata, the older it is. Geologists have divided the strata up into eras, periods, epochs, etc., based upon the animal fossils found in them. For example, we have the Paleozoic, Mesozoic, and Cenozoic eras, which represent the last half-billion years. Eras are divided into periods, and periods into epochs. Once the discovery of radioactivity was made, techniques for dating these ancient strata were possible. Radioisotopes decompose into different elements at a predictable rate. The half-life of a radioisotope is the time it takes for half of the original radioisotope to disappear. We now know the age of each of these geological periods.

Erosion is a process where strata are removed (i.e., eroded away). In northern Utah, you can visit Dinosaur National Park. At this site, well over 65 million years' worth of strata has been eroded away, unveiling fossilized skeletons of giant dinosaurs. In other words, the ground you stand on there is well over 65 million years old. I mention that particular time because scientists see evidence of a global mass extinction at 65 million years ago. It is the so-called K/T extinction, where the dinosaurs vanished. Fossils of dinosaurs, sea reptiles, and pterosaurs can be found in the strata lower than this K/T strata level, but disappear completely in the newer, higher strata.

Many of these fossil skeletons in Dinosaur National Park are completely intact, with the bones articulated (i.e., still connected to each other.) Now, just as deposition can add to sedimentation layers, we know that erosive forces can remove layers. If you have ever stood on the rim of the Grand Canyon in Arizona, you can see how erosion has whittled away layers on some peaks, while leaving them remaining on other peaks. The erosive force of the Colorado River has carved this spectacular landscape over millions of years. It has exposed layers that are dated to over 500 million years old. Near the bottom of the canyon, the layers contain no fossil remains at all. Microscopic life and bacteria existed, but no skeletal life yet evolved.

1.5 STRATA CONTAINING FOSSILS OF HUMAN ANCESTORS

In this book, we are mainly concerned with what the strata can tell us about human origins. From a geological-time standpoint, those strata containing our ancestors' fossils are very recent strata. Although the Earth is 4.5 billion years old, mammals didn't become dominant until the dinosaurs were wiped out by a global mass extinction 65 million years ago. Pre-human fossils may be as old as 6–7 million years, but most of them found to date are far more recent than that. So, on one hand, tens of million years is an incredibly long time relative to human history, but it is a short time span relative to what came before humans.

The human evolutionary story is still being learned. There are startling new discoveries in the field of human origins all the time. Some of the newer finds have shaken the preconceived notions about how the human origin story should progress. Despite the surprises, there are some relatively safe generalizations we can make from what has been learned. For example, I think it would be a totally earth-shaking discovery if someone found a 4-million-years old pre-human fossil outside of Africa. It has become an accepted fact to most scientists that human evolution began in Africa and that we are descended from an African ape ancestry. Fossil evidence suggests that our apelike ancestors never left Africa, but after 2 million years ago, our Homo lineage ancestors migrated out of Africa and into Europe and Asia. *Homo erectus* fossils have been found in all three continents.

Sometimes one can find pre-human fossils by spotting them on the ground. The famous Lucy fossil skeleton, dated at 3.2 million years old, was found in fragments in the rocky ground of Ethiopia. Erosive forces are strongly at work in that region, and new fossils of that ancient time are being exposed continually. Unfortunately, most of them will erode to dust without ever being recognized as fossils. Establishing the age of a fossil is not always feasible, yet often the stratum containing the fossils can be dated using radioisotope analysis. So searching the ground for ancestral fossils is one way they are discovered. Another way is that fossils are discovered in caves. The latter type is harder to date because radioactive minerals are often absent, although extinct animal fossils found in the same cave layer can help establish an age.

1.6   SUMMARY

Seashell fossils on a mountaintop, fossilized trees in the Arizona desert, and dinosaur skeletons in Utah are the parts of our national heritage. They are a part of a geological story of life, told in fossils and strata, that reveals the evolution of life over the last half billion years. Human evolution is a much more recent thing. Human ancestral fossils may extend back to 6–7 million years, but the richer finds are less than 4 million years.

CHAPTER 2

TIME DIVISIONS

CONTENTS

2.1 SCOPE

How did humans appear upon the Earth? For that matter, how did any living thing get here? For most of the Earth's 4.5-billion-years history, nothing larger than one-celled animals existed. In the last 500 million years, multi-celled animals came into being and left fossils behind. The animals existing today have descended from those early ancestors. In this chapter, we will discuss how science has unraveled that story.

2.2 THE HISTORY OF LIFE ON EARTH IS PRESERVED FOR US TO SEE

Geologists have studied sedimentary rock layers from all over the world. They have associated the individual layers with the unique fossil animals and plants found in them. Although the sequence of layers (a.k.a. strata) is found in the same top-to-bottom order throughout the world, recognizing that order from place-to-place is not as simple as it sounds. The sequences may be eroded, tilted, or even inverted by geological forces. Strata may appear different from place-to-place. Finally, the strata of any specific area are incomplete compared to the long geological history of the Earth.

Earth scientists began an effort to correlate similar strata from different regions of the world in the late 18th century. They also formulated a description of its time sequence (i.e., chronology). Abraham Werner (1749–1817) was one of the prominent leaders of this effort. He has been called the father of German Geology for creating enthusiasm in his disciples. The scientists divided the strata into four chronological sections, which they named: primary, secondary, tertiary, and quaternary. We have a different naming system today, but the old one is often referred to in the literature.

The scientists of the 18th century fell into two camps: those who thought the strata was the result of the Biblical great flood and those who thought that the Earth's hot interior was the source of new rock. James Hutton represented the latter school of thought. He presented a paper expounding his interpretation to the Royal Society of Edinburgh in 1785. By the early nineteenth century, the scientific community was focused on identifying particular strata by the unique kinds of fossils found in them. This became the criteria for identifying a particular stratum anywhere on the globe. If two strata contained the same fossils, then those strata were from the same time in history when those particular animals lived. British geologists dominated this correlation and identification process and were responsible for naming the different periods.

2.3   MAJOR TIME DIVISIONS

2.3.1   THE TIME BEFORE VISIBLE FOSSILS

Our planet, like the rest of the solar system, is 4.5 billion years old. The early solar system was a time of great turmoil although there were no living things around at the time to observe it. Hundreds of planetoids, meteors, and comets were on irregular orbits with collisions happening on a regular basis. Some collisions resulted in fragmentations, some in slinging bodies deep into space, and some in building bigger planets. Eventually, the turmoil slowed down and we ended up with eight major planets in stable orbits. However, the planets were far too hot for life to form on them for millions of years. When it did form, it was single cell life-forms, like bacteria. For life forms like plants and animals to exist, multicellular organisms had to evolve first. And multicellular life forms need oxygen to derive their energy requirements. The early Earth had little to no oxygen. It only started to develop an oxygen atmosphere after cyanobacteria evolved and used photosynthesis to build organic molecules. Oxygen was generated as waste. Only in the recent times has the oxygen level has reached adequate levels for multicellular life to exist.

The first visible fossils came from the Cambrian period, some 540 million years ago. The Earth was about four billion years old then. This period is also called the Cambrian explosion because the strange-looking sea floor creatures were amazingly diverse. The discovery seemed to defy the evolutionary theory, which predicted slow gradual change from primitive to complex. Here, bizarre, complex creatures seem to spring into being with no detectable predecessors. The predecessors were eventually found but many of them were too small to be seen without magnification.

2.3.2   THE RELATEDNESS OF LIVING THINGS

Charles Darwin was a very insightful person. He believed that all living things have a common ancestor. He concluded that in the nineteenth century many things we know today were still mysteries to people of his day. For example, it was not understood how the traits of the mother and father are passed to their offspring. Today, we know Darwin's intuition was correct. All life contains DNA and all life is related. Scientists talk about LUCA as the common ancestor of all living things. LUCA is shorthand for the Last Universal Common Ancestor. Somewhere back in time we are related to a mouse, or a fish, or a banana, or a slime mold. The science of DNA even allows us to estimate how long ago each of those common ancestors existed.

2.3.3   THE PALEOZOIC, MESOZOIC, AND CENOZOIC ERAS

Table 2.1 summarizes the geological time line for complex life that left identifiable fossils.

TABLE 2.1   Major Geological Time Divisions for Complex Life Forms

Era	Period	Start Date	Events
Cenozoic	Quaternary	2.6 mya*	Man evolves
	Neogene	23.0 mya	Apes evolve
	Paleogene	66 mya	Small mammals diversify
Mesozoic	Cretaceous	145 mya	New dinosaurs and mammals evolve, K/T mass extinction
	Jurassic	201 mya	Dinosaurs, first birds
	Triassic	252 mya	First dinosaurs, pterosaurs, sea reptiles
Paleozoic	Permian	299 mya	Reptiles diversify
			Mass extinction ends Permian period
	Carboniferous	359 mya	Amphibians and winged insects diversify
			Large trees, first land vertebrates
	Devonian	419 mya	Jawed fishes dominate
	Silurian	444 mya	First jawed fishes and green land plants
	Ordovician	485 mya	Invertebrates diversify
	Cambrian	541 mya	First chordates, diverse fossils

*mya - million year ago.

The *Paleozoic Era* begins with the Cambrian Period with its primitive sea floor creatures and builds toward ever more complex plants and animals. Life was confined to the sea for quite a while and the land masses did not even contain plants as we know them. The early periods were concerned with the evolution and diversification of fishes, trilobites, shellfish, and crustaceans and the evolution of plants. The plants began using lignin, which allowed them to grow larger. Some fish had lungs as well as gills. They were the first vertebrates to occupy land. Lungfish led to amphibians, but they could only reproduce in water. The evolution of the amniotic egg made it possible to break that dependence on water. For the first time, land animals were freed from the restraint of having to lay their eggs in water to be able to reproduce. These fully land-adapted animals diverged into reptiles and mammal-like reptiles. Plants had their great boom in the Carboniferous Period. Moreover, their remains accumulated to become the coal deposits, as we see in Appalachian Mountains. The decaying processes were much slower then because attacking organisms had yet to evolve. Then, the Paleozoic era ended with the largest ever global mass extinction.

The *Mesozoic Era* is also called the age of dinosaurs. Dinosaurs did dominate the land, but pterosaurs came to dominate the air, and marine reptiles came to dominate the sea. Pterosaurs originally had teeth and long tails but lost them over time. Their wings were the membranes of skin stretching from ankle to their extended fingers. These flying reptiles were warm-blooded and had furry coats. Some reached the size of a small air-plane. The sea-going reptiles grew to nearly sixty feet in length and had streamlined bodies and powerful tails. Some resembled the long-necked sea serpents of mythology.

The dinosaurs didn't appear until the latter part of the Triassic, the first period of the Mesozoic era. It took millions of years to recover from the devastating extinction that ended the Paleozoic era. An important thing to know about dinosaurs is that they had a huge advantage over the typical lizard-shaped reptiles in having their legs positioned directly under their bodies. Unlike lizards, dinosaurs did not have to walk in an S-shaped path and that meant they had both lungs working at all times. Moreover, their legs could grow longer, which meant that they could cover more ground. Dinosaurs diversified in the next period, the Jurassic. As one example, huge long-necked dinosaurs evolved during this period. In the final period, the Cretaceous, dinosaurs became more adapted to lower-lying vegetation. Triceratops would be an example. The periods of the Mesozoic era are typified by different climates, plants, and types of dinosaurs. Throughout the era, mammals were diversifying too. The mammals back then avoided the dinosaurs by being nocturnal and staying small. Birds evolved from feathered dinosaurs in this era too. Sixty-five million years ago, the dino-saurs, pterosaurs, and sea reptiles vanished in the global mass extinction known as the K/T extinction. It will be discussed in more detail in the next chapter.

The *Cenozoic Era* is where our ancestor's story really takes off. With the dinosaurs no longer a threat, mammals and birds became the new dominant life forms. Primates evolved from more primitive tree-dwelling mammals and filled the empty ecological niches created by the K/T extinction. Our ancestry traces back to one of these early primates. The more detailed story is told in the next chapter.

2.4   WHERE DO HUMANS FIT IN THIS CHRONOLOGY?

Humans are mammals and although mammals were evolving during the Mesozoic era, the dinosaur threat had kept them small and nocturnal. The situation at the beginning of the Cenozoic era was a world free of the dinosaur dominance and suffering from a dearth of plants and animals due to the K/T extinction event. What was tragedy for the dinosaurs was opportunity for mammals and birds, for they radiated into thousands of new species. The ancestral species to humans are the early primates and they soon appeared to dominate the world of trees.

2.5   SUMMARY

Strata all over the Earth has been correlated chronologically and by the plant and animal fossils found in those strata. The following three major divisions containing fossils have been identified. The Paleozoic era, where life transitioned from primitive sea life to fishes, then amphibians occupying the land, and reptiles as better land-adapted animals. The Mesozoic era followed and is also known as the age of the dinosaurs. A mass extinction killed the dinosaurs and led to the Cenozoic era. Birds and mammals replaced the dinosaurs. We humans appeared late in this geological story.

CHAPTER 3

THE K/T EXTINCTIONS AND THE MAMMALIAN SPECIES RADIATION

CONTENTS

3.1 SCOPE

In Chapter 2, we saw that life has existed on our planet for a long time before we humans ever appeared. We learned the three eras that are so evident in the geological record. In this chapter, we will discuss how the Mesozoic era ended and transitioned into the Cenozoic. The cause of the K/T Extinction is of great scientific interest and it has been extensively explored due to the enormous controversy generated between the meteor impact crowd and the volcanism crowd. We also examine the evolutionary trail of mammals as they became dominant. We are following a fossil trail, which is gradually getting ever closer to humans and their story.

3.2   THE K/T EXTINCTION

Our topic is human origins, so our story becomes closer to home after a landmark event, which happened 65 million years ago. This date is important because it marks the end of the dinosaur reign and opens the door for the adaptive radiation of the mammal and bird species. In geological terms, the Mesozoic Era had ended due to this catastrophic event and the Cenozoic Era began. Dinosaurs, pterosaurs, and giant marine reptiles had dominated land, air, and sea of the Mesozoic Era, but their reign came to an end due to a global extinction event, known as the K/T extinction event. The "K" stands for the Cretaceous period and "T" stands for the Tertiary period (old terminology). It was a major event, and the Cretaceous period was the final period in the Mesozoic Era, and the Tertiary period is the first period in the Cenozoic Era, the era in which we live.

Evidence of the mass extinction is found worldwide, where strata of this age can be examined. Fossil evidence of life is evident below and above the K/T boundary, but a lifeless zone is witnessed at the boundary. The cause of the catastrophe was a combination of a large meteorite striking the Earth and massive volcanism in India. Almost all large vertebrates went extinct as a result of the event. Most plankton, tropical invertebrates, and many plants species also went extinct. Some small vertebrates and other plants and animals did survive the event. They were instrumental in repopulating the Earth.

There have been several global mass extinctions prior to the K/T event, and in each of them, life eventually bounced back. The survivors of the mass extinctions and their offspring carry the flame of life forward until conditions allow the Earth to be replenished. And when that happens, it is a dynamic event, which we call adaptive radiation. There are so few animals and so many available resources for them, that population increases unchecked for some time. The term "radiation" refers to the evolution of new species that occurs under such unique events.

3.3 THE METEOR IMPACT

We know about the meteor impact due to the brilliant detective work of Walter Alvarez and his colleagues. The story is convincingly told in the book *Night Comes to the Cretaceous* by James Laurence Powell. Alvarez and team chemically analyzed the material in the lifeless K/T boundary and discovered that it held an unusually high level of iridium. They sampled the boundary layer at 100 places around the Earth and consistently found high levels of iridium. It is significant that iridium was missing from the adjacent strata. In other words, high iridium levels were only found in the K/T boundary layers. In 1980, the Alvarez team proposed that the iridium had been spread worldwide when an asteroid struck the Earth and sent up a dust cloud containing the rare element. Calculations indicated that the asteroid must have been 10 km (6 miles) in diameter to cause such a dust cloud. Computer modeling predicted that the impact of such a massive body would leave an impact crater 100 km (60 miles) in diameter. The crater was actually found, and it is the Chicxulub crater in the Yucatan peninsula. The crater was identified to be 65 million years old.

That was a highly condensed version of the impact hypothesis. There is a lot more to the full story including the contributions of a Nobel Laureate father (Luis Alvarez) and evidence of shocked quartz, tektites, debris from tidal waves, and particle size distribution of impact material falling to the Earth. Paleontologists, in general, were very skeptical of the impact hypothesis, and so it was a controversial topic at the time, and a lot of scientific and media attention was directed to it.

3.4 VOLCANISM

Prior to the Alvarez Impact explanation for the K/T extinction, most paleontologists and geologists favored a gradualist extinction hypothesis over a catastrophic one. Volcanism was the favored one. When the impact hypothesis was announced, there was a lot of resistance to it. My own first exposure to the controversy was through reading *The Great Dinosaur Extinction Controversy* by Charles Officer and Jake Page. They laid out a persuasive, factual case for why volcanism is the true explanation.

Until I actually saw the overwhelming evidence for the impact hypothesis, Officer and Jake had me favoring their side of the controversy. However, I eventually switched my mind due to the mountain of evidence for the impact.

3.5   HOT SPOTS

One of the most important geological discoveries of the twentieth century was the existence of plate tectonics and continental drift. Impossible as it may sound, continents actually move about. India was once part of Africa and now it is slamming into Asia. The sea floor is continually generated and then subducted under continents. Hot spots are very active and powerful points of volcanic activity. At certain points on the Earth, these deep plumes of molten magma are forcing their way to the surface. The Hawaiian Islands, for example, were formed from such a plume, and new islands continue to be formed. Sixty-five million years ago, there was massive volcanism occurring in India. Enormous quantities of basalt flooded out in the Deccan Plateau of Western India. The huge lava beds that formed, are called the Deccan Traps. They are huge! Even after erosion claimed most of them, their current volume is one million km^3 (240,000 cubic miles).

The Deccan volcanism was not a sudden event but one that continued for a million years or so. It was especially heavy during the K/T extinction time. Huge quantities of dust and lethal gases were released in the vicinity.

3.6   THE COMBINED IMPACT/VOLCANISM HYPOTHESIS

Scientists today rarely argue over Impact versus Volcanism causes for the mass extinction but recognize that both events contributed to global mass extinctions. The focus now is on trying to understand how these events caused the global mass extinctions. The impact phenomena have been compared with the aftermath of global nuclear war with dust blocking out the sun for decades. In addition, there is the concept of red-hot molten rocks returning to Earth and starting forest fires. It is further possible that the impact of the meteor may have accelerated the volcanism that was already underway in India. Between 60 and 80% of all species disappeared during the K/T extinction.

3.7 MAMMALS BEFORE THE K/T EXTINCTION

3.7.1 Reptilian Roots of Mammals

Reptiles first appeared in the last period of the Paleozoic Era, the Permian. This was over 250 million years ago. One kind of mammal-like reptile would gradually evolve into true mammals, and they are called synapsids. Synapsids are distinguished by the single opening in their skulls behind each eye. Moreover, synapsids were also the first four-legged animals (i.e., tetrapod) to have differentiated teeth (i.e., incisors, canines, premolars, and molars). The Paleozoic Era ended with the largest of all global mass extinctions. This was the Permian Extinction. The next era was the Mesozoic, and its first period is called the Triassic. During the Triassic, the survivors of the Permian extinction had undergone adaptive radiation, where new species formed to fill the empty ecological niches. By the mid-Triassic, recognizable mammal species had already evolved. Mesozoic mammals had to cautiously stay out of the way of the then dominant archosaurs.

3.8 WHAT IS A MAMMAL?

Mammals, unlike most reptiles, are warm-blooded. The mammalian bodies are capable of maintaining a constant temperature unlike the cold-blooded fish and reptiles, whose bodies change temperature with the temperature of their environment. Colder temperatures do not slow the mammal's activity level down as it does the reptile. However, the mammal's activity advantage comes with a price. It must find food more often in order to generate the internal body temperature. The mammals evolved a method to retain much of the body heat that they generate: they insulate themselves with a coating of fur. Some mammals, like the whales who live in water, use an external layer of fat to insulate themselves. However, this is a special adaptation to living in water.

Mammals are also unique in having mammary glands. They are able to suckle their young. Mammals, unlike lizards, have their limbs positioned directly under their bodies. Mammals also have diverse teeth, which include incisors, canines, premolars, and molars.

3.9 THREE KINDS OF MAMMALS

Dinosaurs reproduced by laying eggs. In fact, fossil nests containing dinosaur eggs have been well-documented. The earliest mammals also laid eggs, but they did something new as well; they suckled their young. They are called monotremes and there is fossil evidence of them as early as the Triassic Period, the first period of the Age of Dinosaurs. Today, only the duck-billed platypus and the spiny anteater still exist to represent the Monotreme mammals (see Figure 3.1).

Marsupials, like the opossum, are a type of mammals where the fetus crawls into its mother's pouch and attaches itself to a teat for the rest of its development. Australia, before man occupied it, had fauna, which was exclusively marsupial. It is still populated with kangaroos and wallabies.

Placentals are mammals, which carry their fetus in the uterus to term. The fetus is nourished via the placenta. Most of the animals we encounter

FIGURE 3.1 The duck-billed platypus (Courtesy of Wikimedia Commons. From the Wonderful Paleo Art of Heinrich Harder [1858–1935]).

are placental mammals. Cows, horses, pigs, goats, dogs, and cats are all placentals. Humans are placental animals too.

3.10 ANTIQUITY OF MAMMALS

How far back do mammals go? In 1985, the remains of a tiny furry animal (i.e., *Hadrocodium wui*) was found in China that lived in the Jurassic Period 195 million years ago.

Marsupial-type mammals were abundant and widespread during the Cretaceous, the last period of the Mesozoic. Placentals also appeared in the late Cretaceous Period, as much as 100 million years ago. However, most fossil evidence for placentals is found in the Cenozoic Era, after the K/T extinction. Interestingly, there is fossil evidence for hoofed mammals 85 million years ago. So mammals had diversified long before the K/T extinction.

3.11 THE CENOZOIC ERA

The story of human origins may have been begun with mammalian development in the Mesozoic era but visibly takes shape in the Cenozoic. Mammals, now freed of their dinosaur adversaries, became active not only at night but during the daytime. Previously, they were predominately nocturnal, a time that was safe because the carnivorous dinosaurs were inactive. Table 3.1 shows how scientists have divided the new era into periods and epochs. The Cenozoic era begins with a scene of vast devastation. The planet had sparse plant or animal life due to the effects of the K/T extinction. Those animals still existing have a world free of competitors. They can grow their numbers without restraints.

The *Paleogene period* was the warmest time in the new era with a thermal maximum reached 55.8 million years ago. Climate change is a significant factor in evolution, and the breakup of the supercontinent was underway with new continents adrift. India was on a collision course with the continent of Asia. These global landmass changes caused global climate change as well. Although the era began with heavy forestation, the continual cooling of the era saw the breakup of the forests, the expansions of savannas, and the eventual dominance of grasslands.

TABLE 3.1   Divisions Within the Cenozoic Era

Period	Epoch	Dates	Features
Quaternary	Holocene	12000 ya–now	The age of modern man
	Pleistocene	2.5 mya–12,000	Ice Ages, primitive man
Neogene	Pliocene	5–2.5 mya	Bipedal apes evolved
	Miocene	23–5 mya	Apes evolved, Grass spread
Paleogene	Oligocene	33–23 mya	Monkeys evolve
	Eocene	55–33 mya	Early primates, prosimians
	Paleocene	65–55 mya	Jungles, small mammals, primitive proto-primates

New mammal species were developing in the Paleogene. Among them were dogs, cats, pigs, elephants, small horses, rhinos, and others. Birds were also diversifying. The *Paleogene period* is significant to our human story because our ancestors, the primates, were evolving.

In the *Neogene period*, the continents were getting close to their current positions. India was now pushing so hard against Asia that the Himalaya mountains began to rise, with a profound effect on regional climate change. The Neogene also saw the flourishing of bovids (cattle, sheep, goats, antelope, and gazelles). Primitive bipedal apes, like the famous Lucy, evolved during this period.

The *Quaternary period* saw the species radiation of bipedal apes in Africa. They had a brain size similar to that of a chimp but had evolved a skeletal structure designed for walking. One of these Australopithecine species was ancestral to the Homo lineage. The genus Homo is assigned to the fossil men who were taller, bigger brained, and thought to be directly ancestral to us.

3.12   THE AGE OF MAMMALS AND BIRDS

The Cenozoic is also called the Age of Mammals because mammals grew in size and diversity to become the dominant animals of the Earth. Elephants, rhinos, buffalo, and other large mammals filled the large animal niches vacated by the dinosaurs. Whales and other sea mammals, once land animals, evolved to fill the niches vacated by the extinct sea lizards.

Bats evolved to fill the nocturnal opportunities for flying animals. Birds, which had evolved from feathered dinosaurs, have also done an impressive job of radiating into the empty ecological niches. Pelicans, seagulls, and terns specialize in ocean food. Geese, ducks, and teal specialize in freshwater food. Cranes and herons wade through freshwater for tasty bits. Ostriches, rheas, and emus gave up flight to occupy a large animal niche on the land. Pheasants and grouse can fly but prefer running. Penguins gave up flight to be better swimmers. Vultures, ravens, and crows seek carrion for dinner. There are too many songbirds to name, and some birds kill mammals. Eagles, hawks, and owls attack rodents and other prey. There are thousands of both mammal and bird species living today. The radiation of mammals and birds in the Cenozoic is a fascinating story in itself. However, we are focused on the ancestral path leading to humans. The mammalian branch leading to us is the primates.

3.13 SUMMARY

Geological evidence indicated that a global mass extinction, dubbed the "K/T extinction," happened 65 million years ago. However, for quite a while its actual cause was unknown. Massive volcanism was known to occur during this time, so it was deemed to be the likely cause. Walter Alvarez saw it differently. He believed that a massive meteor struck the Earth and caused the extinction and he had evidence to prove it. Iridium, associated with meteors, was found at unusually high levels wherever the K/T boundary was sampled and tested. The point of impact was even identified. Today, it is thought that both impact and volcanism caused the mass extinctions.

The Cenozoic era is often referred to as the age of mammals. Actually, mammals had evolved and advanced an effective reproductive mechanism during the Mesozoic era. Monotreme, marsupial, and placental mammals still exist today but placentals, like us, have become dominant. The Cenozoic era has been broken down into periods, and those broken down into epochs. Monkeys first evolved in the Oligocene epoch, apes in the Miocene, and bipedal apes in the Pliocene. The lineage leading to our species evolved in the Pleistocene epoch. Our species became dominant on the planet in the current epoch, the Holocene.

CHAPTER 4

PRIMATES AND APES

CONTENTS

4.1   SCOPE

Now we zero in on a type of animal more like us. That is, the primates in a general sense and the apes in a more specific sense. These animals have evolved to be better and better at life in the trees. This is a three-dimensional world where stereoscopic vision and a keen sense of balance are highly important. The existing apes are highlighted in this chapter in order to show that more than one type of culture could be successful.

4.2   PRIMATES

The characteristics of primates are based on an adaptation to living in trees. The most primitive primates were squirrel-like, but over time they gradually acquired the characteristics of true prosimians. Today's prosimians include lemurs, lorises, and tarsiers. They have foxlike faces with large eyes, forward-looking heads, and grasping hands. Prosimians are found in the Paleocene (65–55 mya), but it was during the Eocene (55–33 mya) that these characteristics became more fully evolved. Eyes moved to the front of the face to enhance stereoscopic vision and the better depth perception it affords. The foramen magnum (i.e., where the spine attaches to the skull) moved to raise the head to a forward-looking position. Brains became larger. There was an adaptive radiation of prosimians in the Eocene with numerous new species. However, that trend reversed once monkeys evolved. Prosimians today are found mainly on the island of Madagascar.

In the Oligocene (33–23 mya), monkeys evolved and had an adaptive radiation of their own. As monkey populations increased, prosimians declined. Madagascar is one last refuge for prosimians today. It is an island devoid of monkeys. The apes evolved in the middle Miocene (23–5 mya). The gibbons and orangutans of Asia diverged from the African apes about 12–18 mya. The African apes include the gorillas and chimpanzees. Apes differ from monkeys by being smarter, tailless, and swinging under the branches of trees (i.e., brachiation) rather than running atop the branches. Humans also have that same free rotation in the shoulders, which allows brachiation.

4.3   THE FIRST APES

DNA evidence indicates that apes and Old-World monkeys had a common ancestry, 25–30 mya. However, no fossils older than 20 million years had been found. Now, new specimens of fragments of jaws and teeth have improved the picture, taking ape ancestry back to 25 mya. These specimens are from the Oligocene epoch. The location was in Tanzania, and they are the earliest apes found so far.

4.4   APES

During the Miocene when tropical forests were more plentiful, apes occupied a much larger territory. Today, they are only found where tropical forests still exist. One of the most interesting things about the different apes is how they reproduce and the kind of culture that developed as a result. We shall see a harem type culture, where a dominant male controls his females and their offspring, a pair bond culture, where those apes mate for life, And isolated culture, where the male apes exist in two different sizes, and a promiscuous culture, where female apes have many suitors. Now let us examine the apes of today one by one (Figure 4.1).

4.4.1 GIBBONS

Gibbons are our least close ape relative and are actually in a different family, hylobatidae. We humans and the other apes are in the family hominids or greater apes. The gibbons are the most primitive living apes and are closer to monkeys in many ways. Remember that apes descended from

FIGURE 4.1 The Living Apes of the World (From upper right and clockwise: The Gibbon, The Orangutans, The Gorillas, The Bonobos, and The Chimpanzees. (Image of Bonobos by Rob Bixby (JaxZoo_12-16-12-4579.jpg) [CC BY 2.0 (http://creativecommons. org/licenses/by/2.0)], via Wikimedia Commons). and Bonobos courtesy of Wikimedia Commons). [Chimp photo: https://upload.wikimedia.org/wikipedia/commons/a/a2/ Gombe_Stream_NP_Mutter_und_Kind.jpg].

monkeys at a prior time. Gibbons are the one ape excluded from the great apes due to their smaller size and other differences. They do not make nests like other apes, and the males and females are similar in size. Gibbons live in tropical rain forests in Asia. They are currently found in Bangladesh, India, China, and Indonesia.

Gibbons are incredibly good acrobats. I love to see them perform their leaps and swings when I visit the zoo. It is estimated that gibbons had a common ancestor with the great apes about 17 million years ago. That is a much larger time span than the spans between the other living apes. Gibbons form male-female pairs, which often last a lifetime. However, recent studies indicate that while they are socially monogamous, they are not necessarily sexually monogamous.

4.4.2 ORANGUTANS

Orangutans are currently found only in the rain forests of Borneo and Sumatra. The two locations are homes to the two species: Bornean orangutan and Sumatran orangutan. These species diverged about 400,000 years ago. That would make them more closely related than chimps and bonobos. Orangutans diverged from the African great apes 15.7 to 19.3 million years ago. Orangutans are the most arboreal of the great apes due to the fact that they stay up in the trees most of the time. Tigers and crocodiles are their primary predators and that may explain the orangutan's reluctance to spend much time on the ground.

Adult male orangutans come in two different sizes. The larger version is highly sexually dimorphic (i.e., the large male version is much bigger than the female), whereas the smaller male version is only a little bigger than the female. Specifically, the large male is typically 4 ft. 6 in tall and weighs 165 lbs. By contrast, the female orangutan is typically 3.9 feet tall and weighs 82 lbs. This is a significantly large sexual dimorphism. Females only permit the large version male to mate with them. However, the smaller version male will rape a female if no one is around to stop him.

Orangutans are very intelligent and use tools for a variety of uses. Both the Bornean orangutan and Sumatran orangutan species are endangered, but the Sumatran orangutan is critically endangered.

4.4.3   GORILLAS

Gorillas are ground-dwelling apes and are herbivores. They live in central Africa and divide into two species, Eastern gorilla and Western gorilla. They are the largest living primate in size. Males weigh between 300 and 400 pounds in the wild and up to 600 pounds in captivity. Although chimps are our closest ape relative, gorillas are our next closest ape relative.

Male gorillas are much bigger than the females. The dominant male usually has several females in his troop. He is known to defend his troop with his life against carnivores, humans, or other male gorillas. He maintains his harem of females exclusively. Gorilla males have relatively small testicles compared to chimpanzees. This is a physical indicator of his exclusive sexual access to his females. Male chimps, as we will see, have a more competitive situation due to easier sexual access to females, where bigger testicles are needed to produce more sperm in competition with other males. Gorillas are African apes, are plant eaters, and are ground dwelling. Despite their fierce appearance, they are basically gentle animals.

4.4.4   CHIMPANZEES

A chimp's arms are longer than his legs, and his stretched-out-arms span is 1.5 times greater than his height. The human arms span, by contrast, is roughly the same as a human's height. Chimps live in large social groups of both males and females. Usually the group is led by a dominant male, who maintains order during disputes. The alpha male relies on political skill as well as physical prowess to maintain his leadership position. He is likely to be challenged for it at any time. There is also a hierarchical order for female chimps. Their relative position in the group may be inherited from a high-ranking parent. This social complexity indicates the intelligence of chimps. They are also capable of fabricating and using simple tools. Rocks may be used to break open nuts, and sticks may be used to pull termites out of their mounds. Chimps have been taught to communicate in sign language by humans, but they seem to be limited to simple noun-verb sentences.

Chimps live in social groups of both males and females, where the dominant males and dominant females have open sex with each other but interfere with mating of lower status chimps. Male chimps have unusually large testicles. Scientists believe sexual selection favors this physical development because sperm from one male competes with sperm from other males within a given chimp female. Chimps are African apes, live in trees much of the time, eat a wide variety of foods, including occasional meat. Male chimps can be violently aggressive and territorial.

4.4.5 BONOBOS

Whereas male chimps average 3.9 feet in height, male bonobos are slightly shorter. They are also thinner than chimps and have longer limbs. Both chimps and bonobos are knuckle walkers. That is, they walk on all fours by clenching their fists and support themselves on their knuckles. Both can also walk upright albeit unnaturally. They might walk upright while carrying something with their hands. Bonobos are seen walking upright more often than chimps. Bonobos and common chimps are two species of ape recently descended from a common ancestor. Interestingly, bonobos have a quite different social structure from the common chimp. Females have a higher social status than is observed with chimps. They resolve most social problems with sexual favors, not violence. In addition, the females often have sexual relations with other females by rubbing their genitalia together. Bonobos are African apes, are similar in appearance to chimps, but are far less violent and are not male dominated.

The Congo River poses an impassable barrier between the common chimp and bonobo and that barrier may account for the differences in their cultures. The bonobos live south of the river in the Congo and the common chimp lives north of the river. It is easier to find food south of the river, which may account for the easy-going dispositions of the bonobos. Chimps, north of the river, have to compete harder, and as a result, they tend to be more violent. The split between chimp species happened over a million years ago.

4.5 SUMMARY

Primates are tree-dwelling animals, which include prosimians, monkeys, and apes. Prosimians are found in the Paleocene (65–55 mya) and the Eocene (55–33 mya), but stereoscopic vision and larger brains developed in the Eocene. Prosimians thrived until monkeys evolved in the Oligocene (33–23 mya) and replaced them. The apes evolved in the middle Miocene (23–5 mya). Apes differ from monkeys in that they are tailless, smarter, and swing under the branches of trees (i.e., brachiation) rather than running atop the branches.

Asian apes include the gibbon and the orangutan. African apes include the chimpanzee, bonobo, and gorilla. Each of these ape species has a distinct culture and mating behavior.

PROBLEM SET FOR PART 1

QUESTIONS

Q1.   What do fossil seashells on a New Mexico mountaintop and fossil-
       ized trees lying on the Arizona desert floor or an ammonite fossil
       from the Canadian Rockies tell you about the age of the Earth?
Q2.   What geological principle states that lower strata are older than
       higher ones?
Q3.   In which era and period did the first birds appear in the fossil
       record?
Q4.   What does the K/T extinction have to do with the origin of humans?
Q5.   Mammals can be monotremes, marsupials, or placentals. How
       would you explain the differences to a friend?
Q6.   There are two species of chimpanzee. How would you describe
       their differences?

PART II

BIPEDAL SPECIES

Human evolution is not a fanciful theory. We have evidence in the form of fossilized teeth, skulls, vertebrae, femurs, and other bones to prove that it is real. After 2.5 million years ago, we find the first stone tools and weapons in addition to the skeletal parts. Of course, human evolution is only a small and recent entry in the evolutionary history of living things. That larger story is also recorded in the fossil record.

In Darwin's time, only a few fossils of hominins existed. The term "hominin" refers to those species similar to us. Being bipedal is the strongest indicator of being a hominin because we are the only mammal that regularly walks upright on two legs. That is why this section of the book is titled "bipedal species." Between Darwin's time and now, many new species of hominins have been discovered. The scientists who discover and analyze the fossils of these hominins are called paleoanthropologists. A few of them have made some very important finds and have contributed to our understanding of human evolutionary history. We discuss these outstanding scientists and give them credit.

Before looking at individual species of hominins one at a time, we discuss trends that tie these fossil men together in time. For example, bipedal walking preceded the development of larger brains by millions of years. In other words, some apes underwent an evolution, which made them efficient bipedal walkers, yet they still had the brains of an ape. The ancestral line that led to modern humans and Neanderthals is also discussed.

Finally, we examine the major fossil species of hominids one at a time beginning with the primitive *Ardipithecus ramidus* and finishing with *Homo sapiens*.

CHAPTER 5

FOSSILS AND HUMAN EVOLUTION

CONTENTS

5.1 SCOPE

In this section of the book, we shall discuss some of the fossil evidence that relates to human evolution. This chapter will discuss how fossils form and how they are dated. Radioisotopes and how they are used to date fossils is discussed.

5.2 FOSSIL FORMATION

We know something about plants and animals that lived millions of years ago, because they left fossils as evidence of their existence. The Petrified Forest of Arizona is a vivid example of what fossilized plants can be like because these trees are made of rock and not wood. These trees lived over 200 million years ago during the Triassic Period when dinosaurs first appeared. These trees underwent a process that also preserves animals as fossils. The thing to remember is that these particular trees were protected from the processes that usually destroy dead plants. The same thing is true for fossils of animal life. Mainly, that is disintegration due to decay by oxygen and organisms. In the case of animals, there is a host of creatures eager to feed on dead bodies (e.g., crows, vultures, hyenas, maggots).

The petrified forest trees were protected by sediments, which covered the trees before they could be attacked by microorganisms, termites, and other agents. The decaying process then took place very gradually over a long period, and that allowed the decomposed wood to be replaced by minerals such as silica, calcite, pyrite, etc. In the case of Arizona's petrified wood, the different colors resulted from differently colored minerals to replace the decayed wood at different times. The trees were finally exposed as years of wind and rain eroded away the ground over them.

Animal fossils form in a similar manner to plants except it is usually only the animal bones that are preserved. Figure 5.1 is a photo of a dinosaur skeleton taken at Dinosaur National Monument in Utah. Paleontologists painstakingly remove the bones from the rock surface in which they are embedded. Then the bones are assembled into their place in the dinosaur skeleton. Visitors to Dinosaur National Monument are allowed to watch the paleontologists at work.

FIGURE 5.1   Fossilized dinosaur skeleton (Courtesy of Dinosaur National Monument).

5.3   FOSSIL DATING

5.3.1   CARBON DATING

When scientists study the evolutionary history of living things, not only is it valuable to have well-preserved fossils, but it is vital to have as accurate an idea as possible of each fossil's age. For a long time, the techniques for determining a fossil's age were crude or non-existent. However, excellent techniques based on radioisotopes have been developed during the twentieth century. The most commonly known method is radiocarbon dating. Many elements exist in more than one form. The element carbon exists in three isotopic forms namely, ^{12}C, ^{13}C, and ^{14}C. For purposes of the dating test, only ^{12}C and ^{14}C are important. Carbon-12 is most abundant, whereas Carbon-14 is rare. Moreover, Carbon-14 is radioactive with a half-life of about 5730 years. In other words, if we started with 100 grams of Carbon-14, only 50 grams would still exist after 5730 years. A radioactive material is considered totally gone after 10 half-lives. Therefore, the radiocarbon test has no value for dating anything older than say 50,000 years. Other radioactive elements can reach farther back in time.

The radiocarbon test is applicable only to dating objects that contain carbon. Charcoal from an ancient campfire would be an example. The carbon in ancient bones is another. The date that we obtain from radiocarbon dating is the time between now and the date that the animal or plant died. Living things maintain the same ^{12}C to ^{14}C ratio as exists in the atmosphere, and we have determined those values from living things. However, when plants or animals die, the Carbon-14 is no longer replenished and what exists then begins to diminish at a predictable rate.

5.3.2   POTASSIUM-ARGON DATING

Potassium-argon dating is another available dating method. The technique is used to date volcanic rocks. Essentially, the technique determines when the volcanic rock cooled from a molten state. The element potassium is designated by the symbol "K." The radioisotope of potassium, ^{40}K, has a half-life of 1.3 billion years. The ^{40}K decays to form ^{40}Ar. Argon, which is an inert gas, becomes trapped within the rock. By measuring the amount of argon, chemists can estimate the age of the volcanic rock. Sometimes volcanic rock layers are found in strata above and below fossils. While determining the ages of these layers does not date the fossil itself, they do bracket the fossil's age. Other techniques can be combined with this information.

5.3.3 ANIMAL FOSSILS IN THE SAME STRATUM

One of the other techniques is to identify animal fossils in the same layer as human fossils. Paleontologists have amassed considerable data on the periods that different animals have existed. This technique is especially useful when hominin fossils are discovered in limestone caves. There may be no other ways of dating the hominin fossils.

5.4 ADVANCED DATING TECHNIQUES

Numerous new ways of dating fossils have been developed in recent years. They are changing the way paleoanthropologists do their job. More confidence in a fossil's age is possible now that we have several different ways of dating it. If the dates from these different techniques agree, then we feel comfortable with the number. If they do not agree, then they at least provide a range for the fossil's age. Also significant is the fact that the newer techniques use less of the precious fossil material. Among the new techniques are accelerator mass spectroscopy, which greatly reduces the sample size needed; uranium series dating, which can date stalagmites and stalactites; and optically stimulated luminescence and thermoluminescence; which are useful for dating flint, sand, and tooth enamel. For those readers interested in learning more about advanced dating techniques, I recommend the 2012 book, *The Origin of Our Species* by Chris Stringer.

5.5 HOMININ FOSSILS

If Charles Darwin were alive today, he would be amazed at how insightful he was concerning human origins. He had expressed his belief that we had a common ancestor with the African apes and that our evolution originated and took place in Africa. In his time, there were very few fossil hominin intermediates to fortify his belief. If Darwin were alive today, he would be able to examine a copious amount of fossil evidence going back over a span of five million years. Today, the question is not whether there were intermediate ancestors, but why are there so many of them and how are they inter-related. Instead of a clear ancestral chain from ape to human (i.e., a family tree), we are confronted with a family bush.

5.5.1   HOMINIDS AND HOMININS

The terms hominid and hominin are frequently encountered in books on early man, so it is important to know what these terms mean. The term hominid is applied to all living and extinct great apes. That means it includes gorillas, chimps, orangutans, and humans and their extinct ancestors. The term hominin includes living humans and the likely ancestors of humans, but not apes. I think a good test for being a hominin is whether the species normally walked bipedally. Therefore, upright posture and ability to walk naturally on two legs is a major criterion for determining membership. Of course, this bipedal ability has to be verified by skeletal evidence. A humanlike foot with inline toes would be evidence of bipedalism, but toe bones are rarely preserved. Femurs are more commonly preserved than toe bones, and they can indicate bipedalism. Skulls can also indicate that the individual walked upright by how the skull is balanced on the spinal column.

Despite these definitions, many paleoanthropologist authors use the term "hominid" in place of "hominin" in their description of species intermediate between chimps and modern humans. They are not incorrect in doing so, only less precise. Looking at it from a set theory standpoint, hominin members are actually a subset of the hominid set. So, anyone who is a hominin is also a hominid. Some species based on scant fossil evidence are not clearly hominin, but it may be safe to say they are hominid.

Fossilization of ancient hominins is extremely rare. The likelihood of finding those hominin fossils is poor, although these days we have improved techniques, which better our chances. Sometimes the discovery of an ancient hominin only amounts to a tooth or partial jaw. Teeth are the most likely body part to survive over millions of years. Jaws and cranial pieces are next most likely to survive, and partial crania are also all that we have for certain hominin species. Large leg bones are the next likely bone to survive. Mind you, a lot can be deduced from these fragments about the whole individual. Was he human? When did he live? What did he eat? Could he walk upright? These and other questions can be answered by scientists by deductive techniques. However, the confidence level is so much greater when you have ribs, arm bones, spinal sections, leg bones, and if very lucky, fingers and toes.

5.6   WHICH HOMININ SPECIES WILL WE DISCUSS?

This section of the book is going to focus on hominins. The topic deserves an entire book, but we can only devote a few chapters to it. Be aware that it is a dynamic topic with exciting new discoveries happening every few years. We cannot discuss every hominin species ever found, so we will focus on those species where the best fossil evidence exists. I believe it is more meaningful to discuss a hominin species when there are actually fossils of more than one member of that species. I also prefer those hominin species where skeletal bones as well as crania exist.

Similarly, we have to limit our discussion of the hominin discoverers. While there are many good paleoanthropologists, the ones mentioned have made special contributions to the understanding of early man. The Leakey family, in particular, has made enormous contributions.

5.7   SUMMARY

Fossilization is a process where plants or animals are protected from scavenging or rapid decay. This protection could be by burial or other means. As the plant or dead animal slowly decays, the wood or tissue is gradually replaced by minerals until the whole specimen is made of rock. Fossil dating is important to understanding what the fossil represents. Different radioisotopes have different half-lives, so the right one can be selected to date recent or ancient fossils. The coming chapters describe hominid fossils, which may or may not be our ancestors. The hominids selected for discussion have adequate fossil data to properly describe the species, and they are the ones important for the book.

CHAPTER 6

THE PALEOANTHROPOLOGISTS

CONTENTS

6.1   SCOPE

Although there are many paleoanthropologists who could be mentioned, this discussion will be limited to just a few of the more prominent ones. Biographies and/or autobiographies of these pioneering scientists exist, and these books are worth reading. These pioneers have endured hardships to bring the story of human origins to us, and we should not underestimate their contributions. It is impossible to discuss these fossil-hunters without mentioning their discoveries, which were often new species of hominins. The species names follow an international standard. The standard convention for naming a fossil creature consists of two words: genus followed by species. For example, we are *Homo sapiens*.

6.2   THE LEAKEY FAMILY

The Leakey family has made huge contributions to the discovery of our human origins for three generations, and Dr. Louise Leakey is carrying on

the tradition even today. However, it all started with her famous grandfather, Louis B. Leakey (7 August 1903–1 October 1972). Louis was born
in Kenya, a pentagonal-shaped country in northeastern Africa. He was the
son of British missionaries but was as much a Kenyan native as he was a
British national. For example, he built and lived in his own Kikuyu hut as
a youth.

Louis had seen stone tools of the ancient man during his youthful
adventures in Kenya and was determined to prove that Africa was the
cradle of humanity to a very skeptical scientific establishment. During the
early part of the twentieth century, European scientists wanted to believe
that man evolved in Europe, or perhaps Asia, but certainly not Africa.
Only a few hominin fossils had been discovered when Louis began his
lifelong endeavor. Neanderthal fossils had appeared in Europe, and *Homo
erectus* fossils had been found in Asia. Raymond Dart had found fossils
of a bipedal ape in South Africa, but the scientific community scoffed at
his claims of it being ancestral to man. Louis Leakey dedicated his life
to proving that man evolved in Africa and was very successful in accom-
plishing that objective.

The life of Louis Leakey was one of doing great things and also doing
some forgettable things. For one, he had an eye for young women. It is
this combination of behavioral traits that makes us see Louis Leakey as
extraordinary but also human. Louis had a vision that guided his life and
a dedication that is rare and admirable. I highly recommend the biography
Ancestral Passions by Virginia Morrell to learn about his life. Louis was
a prolific author as well as an anthropologist. He wrote and published seventeen books between 1931 and 1977. He also trained and found funding
to study primates in the field for Jane Goodall (chimpanzees), Dian Fossey
(gorillas), and Birute Galdigas (orangutans).

Louis didn't do it all on his own. He certainly wouldn't have achieved
as much without his dedicated wife and partner, Mary Leakey (6 February 1913–9 December 1996). Mary did much of the detailed fieldwork in
the search for early man while Louis wrote books, lectured, and raised
funds, which allowed them to continue their work in paleoanthropology.
Mary Leakey, without formal education, has gained a respected position
as a paleoanthropologist. She has discovered significant fossils including
the first Proconsul skull (a 25 million old ape), *Australopithecus boisei*,

3.75-million-years-old *Australopithecus afarensis* fossils at Laetoli, and 3-million-years-old human footprints at Laetoli.

Of their three sons, Richard was the one to carry on the family tradition. Although Richard vowed not to follow in his father's footsteps and his relationship with his father was contentious much of the time, Richard became one of the most renowned anthropologists and authors of his time. I have read several of his books, learned a lot from them, and have an admiration for him. I particularly liked *Origins Reconsidered*, which is a delightful read and makes paleoanthropology fun to learn. Like his father, Richard had a wife, Meave, who worked in the field with him and carried on the family tradition. Meave is credited with the discovery of the 4.1-million-years old species *Australopithecus anamensis*. Richard also had his young children involved in excavations. Both Meave Leakey and Mary Leakey made a name for themselves as anthropologists. Now Louise, daughter of Richard and Meave, is stepping forward as her parents move toward retirement.

6.3   DONALD JOHANSON

Donald Johanson may be America's most imminent paleoanthropologist. He was born in Chicago in 1943 to Swedish immigrant parents but later moved with his widowed mother to Hartford, Connecticut. He began college with a chemistry degree in mind but later switched to anthropology. He focused on chimpanzee dentition as his doctoral thesis. He made his fame with the discovery of Lucy, a 3.2-million-years old bipedal hominin found in Hadar, Ethiopia, in 1974. She and the other fossilized individuals found near the same site constitute a new species, called *Australopithecus afarensis*. Lucy and her family represent a point in human evolution where bipedal walking was being mastered, and the pelvic bones, legs and feet had adapted to this mode of locomotion. The price for better walking ability was reduced climbing ability in the trees, although Lucy's curved fingers indicated she still did some climbing. Johanson tells the story of her discovery in a compelling book, namely *Lucy, the Beginnings of Mankind*. I highly recommend the book to you. You get to experience the fossil hunter's life in intimate detail. More about Lucy and her species, *Australo-pithecus afarensis,* are presented in Chapter 9 of this book.

Don Johanson is about Richard's age, and they once were close friends. In fact, Don was close to most of the members of the Leakey family and team. In later years, disputes over naming of fossils and other slights damaged that friendship. The story of this feud has been told from both the perspective of Don Johanson and the perspective of Richard in their respective books. I mention this dispute to illustrate the intensity of personal involvement and commitment that paleoanthropologists have for their work.

6.4   TIM WHITE

Tim White is an American paleoanthropologist and a professor at the University of California at Berkeley. He worked with Richard Leakey at Koobi Fori, Kenya, in 1974 and later with Mary Leakey at Laetoli, where she discovered fossil footprints of hominins. Tim White discovered a 4.4-million-years old hominid near the Awash River in Ethiopia in 1994. The nearly complete female skeleton was deemed to be a new species, namely *Ardipithecus ramidus.* A detailed description of this interesting species is provided in Chapter 8. Tim White also discovered an extinct subspecies of *Homo sapiens* in Ethiopia in 1997. The fossils of three individuals were dated at 154,000 to 160,000 years old. The subspecies is called *Homo sapiens Idaltu.*

6.5   LEE BERGER

Professor Lee Berger is a paleoanthropologist with the University of the Witwatersrand, in Johannesburg, South Africa. He has advocated that human evolution took place to high degree in South Africa, although most paleoanthropologists have concentrated their efforts in northeastern Africa with considerable success. Actually, South Africa was the site of the first hominin discovery in South Africa by Raymond Dart in 1924 of *Australo-pithecus africanus*, a fossil commonly known as the Taung baby.

One significant success for Lee Berger was the discovery of a fossil hominid dubbed *Australopithecus sediba*. On 15 August 2008, Lee and his son, Matthew, explored near the Malapa cave and found the clavicle of an extinct hominin. The rest of the skeleton and that of another individual were found nearby. *Australopithecus sediba* had a cranial capacity of *420–450 cc,* which is apelike, yet he had long legs, which is manlike. Additional individuals have been found at the site, which will reinforce and add to the description of this new hominin species. A fuller description of *Australopithecus sediba* is presented in Chapter 10.

Berger topped this important find with his discovery of the new species *Homo naledi* in the Dinaledi chamber of the Rising Star Cave located in the Cradle of Humankind, South Africa. PBS aired a NOVA documentary titled, "Dawn of Humanity," which tells the story of the discovery of the treasure trove of hominin fossils and their rigorous extraction using six slim female scientists, recruited from around the world using a Facebook advertisement from Lee Berger. Numerous individuals were represented by the bones, and the National-Geographic-funded excavation was limited to a three-week effort. Yet a new hominin species having a bizarre combination of apelike and humanlike characteristics was identified. A fuller description of *Homo naledi* is presented in Chapter 11.

6.6 SUMMARY

Louis Leakey was an important paleoanthropologist who not only left a wealth of discoveries for us to ponder but also inspired a family of fossil finders to follow in his footsteps. His wife Mary became a famous discoverer in her own right, his son Richard matched or exceeded his father in great finds, and Richard's wife and children continue the family mission. The nearly complete skeleton of "Lucy" made Donald Johanson famous, but she is far from his only contribution. The even older skeleton of "Ardi" made Tim White famous, but he has a lifetime of contributions besides this great find. Lee Berger has become the latest center of attention with his discovery of *Homo naledi* in South Africa. He may have discovered the greatest treasure trove of hominid fossils ever.

CHAPTER 7

OVERVIEW OF OUR DISTANT ANCESTORS

CONTENTS

7.1   SCOPE

In this chapter, we shall discuss the chain of descent that connects us with the common ancestor of modern humans and the African apes. Genetic evidence, which will be discussed in later chapters, has us most closely related to the chimpanzees with a common ancestor 5–7 million years ago. Unfortunately, there are no fossils of apes in that time period to tell us what our common ancestor actually looked like. Yet from that time period forward, there are fossils of hominins, which walked upright on two legs. Before we examine the hominin species individually in the following chapters, we first want to look at them collectively in this chapter.

7.2   SPECIES NAMES

The standard convention for naming a fossil creature species consists of two words: its genus name followed by its species name. For example, we belong to the species *Homo sapiens*. Lucy belonged to a species named *Australopithecus afarensis* by Don Johanson, her discoverer. If two species have the same genus name, they are considered to be similar and to have a recent common ancestor.

Where do modern humans fit in the scheme of thing? The author Jared Diamond considers us humans to be a third species of chimpanzee. DNA evidence showing genetic similarity would lend credence to his assertion. On the other hand, most people consider humans to be both far superior and far different from apes. In fact, we have evolved into a totally different kind of creature than a tree-dwelling ape in the millions of years that have elapsed. So the DNA evidence of ape-human similarity may not be the most important consideration.

What about the so-called missing links between apes and humans? The fact is that paleoanthropology now has compiled the fossil evidence for numerous bipedal creatures spanning the 5–7 million years since we had a common ancestor with a chimp-like creature until the present. The problem is not so much a lack of fossil evidence but how do we make sense of it all? In order to converse about these different creatures, helps it to give them a name and a description. I may not always agree with the current hominin species names but shall use them in this book anyway because it is most convenient to do so.

7.3   THE HOMININ FOSSIL TRAIL

Quite a few fossils representing different species of early man have been found and together they tell a story of human evolution. One analytic method is to arrange the hominin species by time period and see how the line of descent might have occurred, so it is important to be able to establish the age of each species as accurately as possible. The time span for the earliest appearance and the last appearance has been published in the technical literature. These time periods usually were derived using radio-isotope-dating techniques. In some cases, dates had already been established for animal fossils found in the same stratum. Stone tool cultures are sometimes useful for establishing dates too.

7.3.1   THE AUSTRALOPITHECINES

Table 7.1 is a list of several hominin species arranged in chronological order. The oldest hominins are at the bottom of the table and the most recent are at the top. It turns out that the older hominins differed from tree-dwelling apes mainly in the fact that they stood erect and walked bipedally. Their brain size was still similar to that of modern chimps. We know this by measuring the volume of fossil skulls. One method is to pour lead shot into the brain cavity of the skull and then weigh it. The fossil record clearly shows that bipedal walking and upright stance preceded larger brains. That was troubling because that chronology was not what anthropologists had predicted. It came as a surprise to many because they had theorized that bipedalism must have followed brain size expansion. The fossil evidence says they were clearly wrong.

Some people refer to the bipedal apes of this period as ape-men. Paleo-anthropologists generally refer to them by their genus name, Australopith-ecines. Others shorten the word to Australopiths. These ape-men had been evolving into efficient bipedal walkers from at least five million years ago. In their fossil record, we see skeletal changes that made walking easier and more efficient. For example, ape feet are able to grasp tree limbs because of their opposed hallux (i.e., big toe). Hominin feet, by contrast, have evolved to have all five toes facing forward. Also, the pelvis became shorter and wider. The skull, which is attached to the spine at the back in apes, centered itself for an upright posture. The foramen magnum is the hole at the base of the skull where the spine attaches. The position of the foramen magnum in fossil skulls is often used to assign an upright posture to the species. Despite this tendency of australopiths to trade away climbing ability to gain better walking and running ability, they still maintained some climbing ability. They relied on their tree-climbing abilities to seek refuge from predators. We know this because they retained their long arms and curved fingers.

7.3.2   THE HOMO LINEAGE

You will notice from Table 7.1 that the hominins with the genus Homo do not appear until 2.5 million years ago. The Homo lineage members

TABLE 7.1   Chronological Overview of Hominins

Species Name	When Alive, mya	Comments
Homo sapiens	0.15 to now	Us, the only remaining hominin
Homo florensiensis	0.04 to 0.02	Indonesian island, a dwarf *Homo erectus*
Homo neanderthalensis	0.3 to 0.03	Lived in Ice Age Europe and Middle East
Homo heidelbergensis	0.65 to 0.15	Likely ancestor to Neanderthals and us
Homo antecessor	0.85 to 0.65	Atapuerca, Spain, oldest European hominin
Homo erectus	1.8 to 0.3	Lived in Africa, Asia, and Europe
Homo georgicus	1.8	Dmanisi, Georgia, an early *Homo erectus*
Homo ergaster	1.9 to 1.4	An early *Homo erectus*, exclusive to Africa
Homo habilis	2.3 to 2.0	Africa, few specimens, tool user.
Homo rudolfensis	2.5 to 1.9	Kenya, East Africa, single specimen
Paranthropus robustus	2.0 to 1.5	A body like *A. africanus*, but massive face and molars
Paranthropus boisei	2.1 to 1.1	More massive face and molars than *P. robustus*
Paranthropus aethiopicus	2.6 to 2.3	E. Africa, Black skull specimen, massive face
Australopithecus sediba	1.8 to 2.0	Recent find in South Africa dated at 2 mya
Australopithecus afarensis	4.0 to 2.8	East Africa, Lucy & other fossils, bipedal but apelike
Australopithecus garhi	2.6 to 2.5	Africa, first toolmaking hominids known
Australopithecus africanus	2.9 to 2.0	South Africa, rounder, larger skull than *A. afarensis*
Australopithecus bahrelghazali	3.6	Chad, Africa, mandible, may be an *A. afarensis*
Australopithecus anamensis	4.1 to 3.9	Likely ancestor of *A. afarensis*
Ardipithecus ramidus	5.8 to 4.4	Bipedal walker but primitive and apelike

resemble Australopithecines at the beginning and modern humans in the present. So chronologically, the Homo species came to resemble modern humans more and more as time went on over the 2.5-million-years time span. Perhaps the most significant change during this period is a threefold increase in brain size compared to the starting point of 300–500 cc range for apes. With *Homo erectus,* the legs got longer, arms got shorter, and the gut got smaller too. Whereas the ribcage of Australopithecines tends to flare outward to accommodate a big gut, the Homo ribcage became straight up and down. Arms got shorter as the Homo lineage stopped climbing and sleeping in trees. *Homo erectus* tamed fire, developed stone cutting tools, and became reliant on a diet of meat in addition to his plant foods. His face got progressively flatter, teeth smaller, and skull rounder. He, for the first time in the animal kingdom, learned to speak and communicate. He alone cooked his own food. He alone fashioned his own clothing.

7.3.3   THE ROBUST AUSTRALOPITHECINES

You will also notice from Table 7.1 that hominids with the genus *Paranthropus* coexisted with the early Homo lineage hominins. Sometimes you will see these species identified with the genus *Australopithecus* instead of *Paranthropus.* Some paleoanthropologists consider all of these bipedal apes to be in a single genus (*Australopithecus*) and split it into gracile and robust types. It seems that the graciles either went extinct or became the Homo lineage whereas the robust types developed bigger cheek teeth and massive jaws and supporting structure in the skull. Some of them even had a bony ridge on the top of the skull to attach their massive chewing muscles. Apparently, their survival depended on adapting to hard-to-chew food, like roots and tough vegetation. Their coexistence with the Homo lineage in Africa was another surprise for paleoanthropologists. It had become a general rule that two or more similar species cannot long coexist in the same area at the same time. And yet their fossils proved the exception to the rule. Actually, two similar species cannot coexist if they compete for the same kind of food. One species will survive and the other will disappear. Obviously, they were not eating the same kinds of food. Paranthropus was eating raw tough vegetation and Homo was eating cooked meat and vegetation.

7.4   INDIVIDUAL HOMININ SPECIES

In the following chapters of this section, namely Chapters 8–14, we individually discuss some of the more important hominin species. As you read about them, try to construct a bigger picture in your mind of what the evidence might be telling us.

7.5   SUMMARY

Two of the things we shall especially look for as we survey individual hominids are when they lived and how big was their brain. Radioisotope dating can often bracket the fossil's time, and animal fossils can confirm it. Brain size can be determined if adequate skull material has been preserved. We are also interested in what genus has been assigned to the fossil hominid species. Was it an Australopith or of the Homo lineage? If the first genus, then was it gracile or robust? Paleoanthropologists look at fossil characteristics and sometimes reach different conclusions about these points. Finally, one lineage called by the genus Homo was on the evolutionary road to becoming us. The species so designated underwent significant evolutionary changes in the process.

CHAPTER 8

ARDI

CONTENTS

8.1 SCOPE

In this chapter, we shall discuss *Ardipithecus ramidus*, the oldest fossil hominid with a preserved skeleton. Her discoverer, Tim White, decided this species was different enough from the Australopiths to assign a new genus to it. Ardi's foot had not evolved to the point where she had five inline toes like the Australopiths. Ardi is interesting because she was still not an efficient bipedal walker 4.4 million years ago, but she was on her way, compared to her ancestors.

8.2 WHO WAS ARDI?

Ardi is short for *Ardipithecus*, the genus of perhaps the earliest bipedal walker ever found. Ardi was a quite complete, female fossil hominid who lived 4.4 million years ago. She stood 3 ft. 11 in. tall, weighed about 110 lbs., was small-brained and built for tree climbing, yet was also able to walk bipedally. As you will see in the pages ahead, different appearing man-ape fossils were discovered in Africa in terrain dated between 3.5 million and 2 million years old. Most of them were considered to be in the same genus, namely, *Australopithecus*. However, their anatomical differences earned them status as separate species. Tim White deemed Ardi to be too primitive to join that group.

8.3   ARDI'S DISCOVERER

Tim White is a paleoanthropologist and professor of integrative biology at the University of California at Berkeley. White's name keeps popping up around important finds of discovered hominins. In 1974, while working on his PhD at the University of Michigan, White was fortunate in being selected to do summer fieldwork with the imminent Richard Leakey at Koobi Fora, Kenya. He then was assigned to work with Richard's mother, Mary, at Laetoli, Tanzania. They discovered hominin teeth and jaw fragments dated at 3.7 mya. However, their most memorable and remarkable find at Laetoli was a trail of hominin footprints preserved in volcanic ash. Tim White later collaborated with Donald Johanson in examination of Lucy and the first family fossils. White is respected for his directness and accuracy but less for his tactfulness, it is said.

8.4   DESCRIPTION OF *ARDIPITHECUS RAMIDUS*

Ardipithecus ramidus is the name given to a 4.4-million-years-old fossilized individual in 1994 (See Figure 8.1). It was Tim White's team that discovered this female hominid in the Afar depression in Ethiopia in 1992. This region is only 46 miles from the site where the famous Lucy skeleton was found. We shall discuss her in the next chapter. Lucy represents the species *Australopithecus afarensis*, which lived about one million years later.

Seventeen fragments of Ardi were found including skull, mandible, teeth, and arm bones. More fragments were found in 1994, bringing the total to 45% of the skeleton. In 1999, a different team led by Sileshi Semaw spent four years in the Afar and turned up fossils of nine more similar beings. Debate continues whether or not Ardi was a hominin, i.e.,

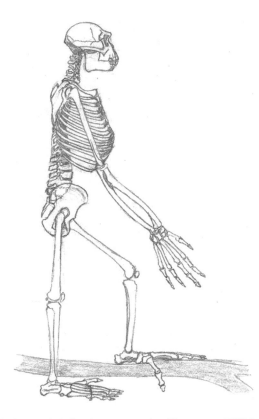

FIGURE 8.1   Skeleton of *Ardipithecus ramidus* (Courtesy of Wikimedia Commons by Tobias Fluegel [own work] [GFDL, http://www.gnu.org/copyleft/fdl.html].

a true ancestor of ours. One feature that set Ardi off from the australo-pithicines was that the hallux (i.e., big toe) was opposed to the other toes. This indicates some retention of an apelike gripping ability of her feet and that she still did a lot of tree climbing. However, compared to the feet of chimpanzees, Ardi's feet were better suited for walking. Moreover, her pelvis is more humanlike than chimp-like. The long arms and long curved fingers are an apelike feature adapted for climbing and for swinging under tree branches. She did not have hands as useful for tool making as we do. Ardi had a small brain (300–350 cc), which is about the same size as the brain of a bonobo.

One feature that shows change is the teeth. The canines are smaller than chimp canines and somewhat apelike, except they wear on the tips, like human canines. By contrast, ape canines hone on the first premolar. The thin enamel on Ardi's teeth suggests that she had a diet of soft foods like fruit, similar to what chimps eat today. Also, Ardi's lower face is not as protruding as in a chimp. Moreover, the smaller canines and equivalent size for males and females suggest that Ardi and her type led a less combative male-to-male type relationship than we observe in modern chimpanzees.

8.5   ARDI'S PLACE IN HUMAN EVOLUTION

Ardi's age of 4.4 million years puts her chronologically closer to the most recent common ancestor of chimp and man than the majority of hominin fossils found. One thing we can deduct from her skeletal features is that she was both a primitive bipedal walker and an adept tree climber. She moved as a quadruped in the trees. Some paleoanthropologists think she was more representative of that common ancestor we have with chimps than modern chimps are. Modern chimps have evolved during the 5 million or so years since that common ancestor. They have specialized in knuckle-walking when moving on the ground.

Ardipithecus ramidus is not the only species in the genus. White also found a species dubbed *Ardipithecus kadabba*, which is dated at 5.6 million years old. This species is based on teeth, and a few bone fragments, and these fossils indicate that kadabba was a more primitive hominid than Ardi.

There are scientific arguments for including Ardi's in the human ancestry line. For example, isotope studies of her tooth enamel revealed that she fed both arborally and in open habitats. The latter would differentiate her from apes. Another study found that Ardi, *Australopithecus afarensis,* and *Australopithecus sediba* all had hand bone features that would have us include them in the human lineage and not in the ape lineage. There are also unique features of the brain organization in Ardi that are found only in Australopiths and the homo lineage.

8.6   SUMMARY

Ardipithecus ramidus is a species of hominid that may or may not be on the ancestral path to us. It is debatable. And yet, Ardi is a very fortunate fossil find because it is very wishful thinking to expect to find a 4.4 million-year-old skeleton with finger and toes still preserved and, yet she does. From the reconstructed foot, we see a creature capable of walking upright, but with quite primitive feet. She still retained a grasping foot, with hallux opposed to the other four toes. Ardi was not a total tree dweller as tooth analysis shows she ate foods from open habitats. She could walk bipedally as the characteristics of her pelvic would so indicate.

CHAPTER 9

LUCY AND *AUSTRALOPITHECUS AFARENSIS*

CONTENTS

9.1   SCOPE

Lucy may possibly be the most famously known bipedal ape ever. She was big news when she was discovered, and she is still one of the most important discoveries in paleoanthropology due to the extraordinary number of bones recovered in her skeleton and her great antiquity. Lucy represents an extinct species known as *Australopithecus afarensis*. This species is well-documented because several other members have been discovered and examined in addition to Lucy. This chapter tells us more about them.

9.2 DONALD JOHANSON

Lucy may be one of the most famous hominin finds ever. Donald Johanson and Maitland Edey tell the story of her discovery in one of the most interesting and compelling books in the field of paleoanthropology, namely *Lucy: The Beginnings of Humankind.* When I read the book, I felt as if I was on the expeditions with them, feeling the frustration when obstacles blocked our path and also feeling the joy when the great discoveries were being made. Interpretation of the fossil evidence was a fascinating process leading to a new understanding of what these creatures really were. Up until the time of her discovery, the best paleoanthropological find was that of a Neanderthal skeleton from 75,000 years ago.

Lucy's discoverer, Donald Johanson, was a budding American paleoanthropologist who had focused on the dentition of chimps for his doctoral thesis. It was a wise career decision because teeth are preserved as fossils better than any other body part and the features of teeth tell us quite a bit about how that creature lived. Johanson was funded to travel to Europe to examine the extensive European chimp teeth collection and on to South Africa to examine their australopithecine skulls and teeth.

9.3 BACKGROUND TO THE DISCOVERY

In 1973, he was given the opportunity to head an American team in a French/American joint effort. The location was Hadar, which is a village on the southern edge of the Afar Depression. Also known as the Afar Triangle, it is the junction of three geological plates. This area is part of the Great Rift Valley. It is essentially a desert, with fossils being continually exposed by erosion. His counterpart in the French team was Maurice Taieb. Older than Johanson, Taieb was an expert in the geology of the Hadar formation. He is both a geologist and a paleoanthropologist. Taieb made extensive surveys of the area, attempting to accurately date the different strata.

Johanson had been worrying about burning through his budget without having much to show for it when he discovered the femur and tibia of ancient hominin. The age of the stratum was dated at about three mya, so this was the first ever find of such antiquity. Moreover, the angle of the knee joint was like that of modern man and unlike that of an ape. This person had walked on two legs.

9.4   DISCOVERY OF LUCY

During his second field trip to Hadar in 1974, Johanson made his most famous find, the petite female man-ape from over three million years ago, known throughout the world thereafter as Lucy. The name "Lucy" was inspired by the then-current playing of the famous Beatle's song, "Lucy in the sky with diamonds." Fossil Lucy was a magnificent find, being so ancient and still having much of her skeleton intact. Although small-brained and having an apelike skull, Lucy was an advanced bipedal walker. Her skeleton proved it (see Figure 9.1.) Lucy was a young adult when she died. We know this because her wisdom teeth were in. She had a *V*-shaped jaw, which is considered primitive. Humans have an oval-shaped arch. Too much of her skull was missing to be able to measure her cranial capacity. Other A. *afarensis* skulls show a more oval-shaped arch differentiating them from apes and gorillas. Moreover, they had larger molars and premolars than apes and thicker enamel. This fact plus the wear marks on the teeth indicate A. afarensis was consuming coarser, more fibrous foods than tree-dwelling chimps.

9.5   THE FIRST FAMILY AND WHAT THEY ARE

During the third field session in Hadar, a National Geographic photographer named David Bell paid them a visit. He had hoped there might be some new discoveries for him to shoot. The next day, he had the greatest shoot of his life occur. At site 333, hominin fossils abounded. Johanson's team was in a state of elation, eagerly searching and finding fossil after fossil. All of the activity was filmed. This was again a first. In the following days, the process became more orderly. Altogether, fossils from 13 different individuals were found. Men, women, and children were in the party. This group of individuals was dubbed the "First Family." Although unknowable, Johanson and team surmised that a drowning accident might have killed them, perhaps from a flash flood.

9.6   THE GIRL FROM DIKKA

In 2006, a team led by Zeresenay Alemseged found another *A. afarensis* individual. This was across the river from Hadar at a site called Dikka.

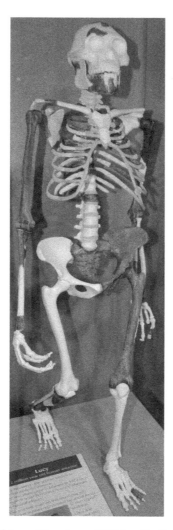

FIGURE 9.1   A replica of Lucy (Courtesy of Wikimedia Commons by Chiswick Chap [own work] [CC BY-SA 3.0 [http://creativecommons.org/licenses/by-sa/3.0].

This was a young girl, about three years old. She was nicknamed "Selam." Much of her skeleton was intact, including bones, which are rarely preserved. These include the scapula (i.e., a shoulder bone) and the hyoid bones (i.e., throat area bone). The accumulating data on the A. afarensis species is making it the most studied species of such antiquity.

9.7 ANALYSIS OF THE FOSSILS

Back in his lab in Cleveland, Johanson was hesitant about interpreting and reporting where his remarkable fossils fit into the scheme of things. He didn't want to make a mistake and be professionally humiliated. He decided that he needed another pair of eyes interpreting along with him and felt that Tim White was best suited to the task.

Johanson and White spent the summer examining the fossils and debating what it meant. Lucy and the Site 333 family had human-like bodies, but their skulls were closer to ape skulls than to humans. Their cranial capacity ranged between 380 and 450 cc's. This was slightly bigger than the range for chimps (i.e., 300–400 cc). After exhaustive analysis, Johanson and White both agreed that the Hadar fossils represented a new species, never discovered before. These man-apes lived between four and three mya. There is no evidence that they used stone tools. They were assigned the name, *Australopithecus afarensis*, after the Afar region in Ethiopia where they were found. Johanson believed that Lucy and her kind were ancestral to the Homo genus, which led to modern man.

9.8 DESCRIPTION OF *AUSTRALOPITHECUS AFARENSIS*

Fossils of *A. afarensis* have been discovered in several areas of Ethiopia and Kenya. These fossils have been dated between 3.2 and 3.4 million years ago. Although clearly capable of walking bipedally, *A. afarensis* retained many apelike characteristics. They had a small brain (about 380–430 cm^3) and had a face with forward-projecting jaws. Some physical characteristics suggest that they climbed trees as well as spending time on the ground: (i) the scapula was more apelike than human, and (ii) the curvature of the fingers and toe bones was similar to the curvature seen in modern apes. Curved fingers and toes are arboreal adaptations. On the other hand, they had lost the grasping ability of the arboreal foot with its adducted big toes. Skeletal adaptations for bipedal walking were also evident: (i) the pelvis was far more human-like than apelike, (ii) the femur angled from the hip inward toward the knee, and (iii) the feet were more human-like than apelike. These three traits showed that *A. afarensis* had undergone considerable skeletal change in becoming an efficient bipedal walker even though they were still at the primitive point in the australopithecine evolution.

Human-like footprints made 3.6 million years ago were preserved in Tanzania. It is more than likely that they were made by *A. afarensis*. This is the famous Laetoli site discovered by Mary Leakey. The reason that paleoanthropologists think they were made by Lucy's kin is that teeth and jaws also discovered at the site seemed to be a close match for the *A. afarensis* fossils from Hadar in Ethiopia. In addition, they had actual foot bones from Hadar to compare with the tracks.

9.9   ANOTHER INTERPRETATION OF WHAT THE FOSSIL EVIDENCE SHOW

Dean Falk emphatically disagrees with Johanson. She does not believe the First Family is comprised of only one species, nor that *Australopithecus afarensis* is ancestral to the Homo genus. Dean Falk, an anthropology professor, authored the book, *Brain Dance*, where in Chapter Six, she presents her arguments. Professor Falk has devoted her career to studying the brain and what it can tell us about human evolution. First of all, she does not believe the geological evidence suggests that the First Family died together. It is more likely that the individuals died separately and were buried at separate times. Dean Falk tells us that Johansen at first believed that three hominin types were represented by the First Family: gracile australopithecines, robust australopithecines, and early Homo. She thinks that Tim White influenced the decision to lump them all under one species label. He thought it likely that robust australopithecines were males and gracile australopithecines were females.

It had been a general rule in anthropology that two species of a similar animal could not coexist at the same time and place. However, in 1976, Richard and Alan Walker proved that the general rule was wrong. They showed that both a *Homo erectus* and a robust australopithecine coexisted. That important finding ended the single species hypothesis.

In 1982, Dean Falk and Glenn Conroy discovered that robust and gracile australopithecines can be distinguished from each other by noticeable grooves located on the backside of the braincase. The groove is called the occipital/margin sinus (O/M for short). It was unusual because it is rarely large enough to leave a groove in living apes and humans. Yet its presence correlated well with robust australopithecines and not at all with gracile australopithecines. Falk and Conroy explained the O/M groove as an early adaptation for blood drainage from the brain for bipedal walkers. The enlarged O/M sinus was a pathway to the vertebral plexus of veins. It is this vein network that serves to drain cranial blood when a person stands up.

They felt that this evidence pointed to *Australopithecus afarensis* being ancestral to robust australopithecines, not the Homo line. In other words, Lucy and her kind could not have been ancestral to us.

9.10   LUCY'S ANCESTORS

A species of hominin older and more primitive than *A. afarensis* is the species called *Australopithecus anamensis*. It is believed to be ancestral to Lucy's clan. There are 78 specimens, mainly dental remains, that identify this four-million-years old creature. Meave Leakey and Alan Walker discovered fossil fragments in 1994, including one complete lower jaw, at a site near Lake Turkana. Fossils have also been found in Ethiopia.

9.11   AN IN-DEPTH ANALYSIS OF THE FOSSIL EVIDENCE

For those of you who would like to see a deeper analysis of what the fossil evidence tells us about ancestry, how many species actually are represented by the first family fossils, and where is *Australopithecus afarensis* in the evolutionary chain to man, I recommend you read Chapter 3 in the book *Extinct Humans* by Ian Tattersall and Jeffrey Schwartz. One thing that you will learn is that there is nothing simple about human origins.

9.12   SUMMARY

Australopithecus afarensis is an important species to the study of early man and much can be learned from these fossils. Its first member was discovered by Donald Johanson in Hadar, Ethiopia in 1974. Lucy was dated at between 3.2 and 3.4 million years old and was not only the oldest hominid discovered at that time, but her skeleton was quite complete. Analysis of her bones showed that she was an accomplished bipedal walker, but her jaw indicated apelike characteristics. Unlike Ardi, who was discovered nearby, Lucy had inline toes and a foot much like our own. Johanson's luck did not end with Lucy because he discovered the first family consisting of 13 individuals. In collaboration with Tim White, Johanson deemed the individuals to be all of the same species as Lucy. Other anthropologists are also not sure whether this is true.

CHAPTER 10

AUSTRALOPITHECUS SEDIBA

CONTENTS

10.1 SCOPE

The important finds in paleoanthropology were coming repeatedly from Northeast Africa (Kenya and Ethiopia) to the point where it seemed like they were the best place for a fossil hunter to really hit pay dirt. However, the place of recent interest is South Africa, thanks to paleoanthropologist Lee Berger. In this chapter, we shall discuss one of his great discoveries; that of *Australopithecus sediba*.

10.2 DEATH TRAP PRESERVES TWO-MILLION-YEARS OLD HOMINIDS

A new species of australopithecine (i.e., man-ape) was discovered by Lee Berger in South Africa in 2010. These hominins are unusual in displaying a combination of both primitive as well as homo-like features. About two million years ago, several such individuals fell to their death in a limestone pit, and parts of their skeletons were preserved as fossils. From an anthropologist's viewpoint, these fossilized bones were a fortuitous find indeed. Bones that are rarely preserved, such as finger bones, are beautifully preserved at this site. The fossil-containing rocks were moved to Professor Berger's laboratory, and the fossils were carefully removed and examined. The skeletal traits of the individuals were found to represent an entirely new species of hominin.

The new species, *Australopithecus sediba*, is described from six skeletons recovered from the Malapa Fossil Site at the Cradle of Humankind World Heritage Site in South Africa. They were a juvenile male, an adult female, and adult male and three infants. Analysis of bone damage indicates that these individuals fell to their death at the bottom of the pit. Other animals that had also fallen to their deaths were found at the bottom of the pit. Those animals include saber-tooth cats, mongooses, and antelopes.

10.3   NINE-YEAR-OLD BOY DISCOVERS IMPORTANT FOSSILS

Paleoanthropologist Lee Berger excavated the fossils from the now-shallower pit. The initial discovery was made by his nine-year-old son, Matthew, in 2008. Berger had been surveying limestone caves in the area for years, and on this occasion, he brought his son and his dog to the site to keep him company. He told Matthew to go find a fossil, and to his surprise, Matthew quickly found a rock with a hominin clavicle sticking out.

When Berger returned again to the site with fellow scientists, they found nothing further for quite a while. Then Berger spotted hominin bones on another rock within the pit. When he lifted the rock, two hominin teeth fell into his hand. It soon became evident that more than one individual was represented by these fossils. It was later determined that two million years ago the pit was much deeper and that these individuals had fallen to their deaths. Analysis of the stress lines in the bones was consistent with a hard fall.

The cave turned out to contain a rare assemblage of skeletal bones. The finger bones, for example, were preserved and in such good condition that small bones of the hands could be reassembled, hands, which by the way, were adapted to tree climbing. Although a tree-climber, *A. sediba* was also a bipedal walker as indicated by his lower anatomy. In addition to his good fortune with skeletal finds, Berger was also fortunate that the site could be accurately dated with radioisotopes. Moreover, animal fossils were discovered at the site to help confirm the two-million-years-old age for the fossils.

10.4 FOSSIL DETAILS

A. sediba would have excelled at tree climbing yet also walked bipedally. Her gait was unusual though. She walked with her shoulders shrugged and with her feet rolling inward with each step. The unusual gait was deduced from examination of the foot and ankle. She had an advanced ankle bone and a human-like arch but had a primitive heel (Figure 10.1).

We even know what kinds of food *A. sediba* ate. She ate fruits and other forest foods. It was possible to identify plant remnants in the plaque of the fossil teeth. The mandible and tooth size are quite gracile, similar to what is found in *Homo erectus*.

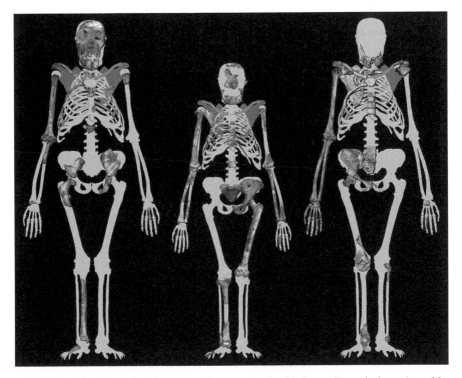

FIGURE 10.1 *Australopithecus sediba* compared with Lucy (Description: *A. sediba* (MH1) – left, Lucy (AL288-1) – center, *A. sediba* (MH2) – right)); (Courtesy of Wikimedia Commons by Peter Schmid courtesy of Lee R. Berger, University of the Witwatersrand.).

A. Sediba had long arms like an ape, so one would expect to see long curved fingers too. Surprisingly, this is not the case. The fingers were short and straight. The long thumb and fingers come together in a precision grip typical of humans. The relative difference in the sizes of *A. sediba* males and females was much like that of modern humans.

Fossil individual MH1 was a young boy of about 12–13 years old, 4 ft. 3 in. tall. The skull of MH1 was complete enough to be able to measure the brain size. The cranial capacity measured 420 cm^3; this would be at the high end of the cranial capacities of *A. africanus,* an earlier hominin from southern Africa. This brain capacity would be too low of a number for the species to be included in the *Homo* genus though yet, there are facial features of the skull that are more human-like than is typical for australopithecines.

10.4.1   POSITION IN THE HOMININ LINE UP

A. sediba is more similar to *A. africanus* than it is to any other hominin. *A. africanus* fossils had been discovered in South Africa by Raymond Dart and others. *A. africanus* lived 2 to 3 million years ago. *A. sediba* is dated at 2 million years old. Lee Berger argues that *A. sediba* has distinct enough features to be given the status of a separate species. Not all paleoanthropologists agree, noting that there might have been a great deal of variation in the *A. africanus* species.

Her numerous humanlike traits could make her transitional between *A. Africanus* and *Homo habilis* or even *Homo erectus.* Lee Berger argues for the find being ancestral to humans, whereas Tim White and others suggest that *A. sediba* was an australopithecine, which co-existed with early members of the Homo lineage.

10.5   SUMMARY

Australopithecus sediba adds a new dimension to the fossil record leading to us. Another kind of Australopith has joined the list. This one has been dated accurately at about 2 million years old, and the completeness of the fossils found are excellent. We not only have six individuals, but we have numerous post-cranial bones. These individuals fell to their deaths in a pit. Analysis of the trauma to their bones proves it. Lee Berger discovered these fossils after his son, Matthews, found a rock containing a hominid bone in it. Although this species is most like *Australopithecus africanus* than any other existing Australopith, Berger has assigned it a separate species identification based on its unique characteristics. It is a mixture of both human and Australopith features.

CHAPTER 11

HOMO NALEDI

CONTENTS

11.1 SCOPE

The discovery of *Homo naledi* is one of the most fascinating stories of paleoanthropology. Hominin fossils had been spotted in a cave in South Africa. However, this cave cannot be not easily accessed. It requires a daring, slim contortionist to traverse its restricted path to the chamber of fossil bones. Lee Berger wanted to harvest the hominin bones that his cave explorers saw and those buried beneath them, but it requires specialists to do it properly. So, he put out a worldwide call for suitably shaped anthropologists to undertake this risky mission. Amazingly, he was swamped with eager and qualified candidates. After assuring for their safety on this arduous descent, the fossil extraction process began, and he was soon overwhelmed with pristine fossil bones. As his funding ran out, this leg of the acquisition journey came to an end. However, the session ended with numerous bones still in the cavern waiting to be retrieved. Such was the experience of Lee Berger, who directed the extraction and is now studying the wealth of fossils in his laboratory. In this chapter, we tell his story and learn what the bones tell us about human evolution.

11.2   THE IMPORTANCE OF *HOMO NALEDI* TO PALEOANTHROPOLOGY

As I write this chapter, the mystery of the *Homo naledi* fossil find is in an early stage of being unraveled. Until *Homo naledi*, the chronology of hominin evolution followed this theme: Apelike creatures took up bipedal walking, yet retained tree-climbing physical traits like long arms and long curved fingers and toes. We call these bipedal apes by the term "australopithecines" or "australopiths." About two million years ago, one group of them (perhaps yet undiscovered), developed larger brains, longer legs, shorter arms, and lost many of their tree-climbing attributes. We call them the Homo lineage. Over time, Homo brains got bigger and bigger and the evidence shows that we modern humans evolved from these early humans.

The current problem is that the *Homo naledi* discovery adds much complexity to this accepted theme because these creatures are composed of an unusual mixture of both Australopith and Homo lineage traits. The skull, dental arch, and teeth are clearly Homo. Although they walked bipedally, their pelvis is flared like australopithecines, and they had curved fingers adapted for tree-climbing. Their brain size was less than what we have come to expect for Homo fossil-men. Finally, they appear to have intentionally buried their dead for a long period of time. This practice was thought to be a distinguishing sign of advanced cultural and human development. And so, for burial to have been practiced by a primitive, small-brained homo lineage species is totally unexpected. We have fossil evidence of intentional burial that is hard to believe. However, it seems impossible to dispute.

So far, 15 individuals have been examined and they all have this strange mixture of primitive and advanced traits. Waiting to be excavated are many more fossil bones in the Rising Star Cave. The wealth of hominin fossils is more than any scientist could ever hope to find, but there is a catch. Until recently, these fossils were deemed to be somewhere between hard to date and impossible to date. However, never doubt the ingenuity of man because they have now been dated at between 236,000 and 335,000 years old.

11.3 DISCOVERY

Rick Hunter and Steven Tucker discovered the cave chamber containing *Homo naledi* while working with Pedro Boshoff. They reported their find to South African paleoanthropologist Lee Berger. In the previous chapter, we saw that he had discovered *Australopithecus sediba* fossils in a different limestone cave. Together, the two cave explorers found one of the greatest caches of hominin fossils the world has ever seen. The cave that paid off so well was the Rising Star Cave system in the Gauteng province of South Africa.

The story of the discovery and the excavation have been made into a documentary titled "The Dawn of Humanity" and aired by NOVA. The chamber, where the fossils were discovered, was difficult to access, to put it mildly. Only very lean and athletic people could possibly do it due to the tight squeezes and strenuous rock climbing involved.

11.4 EXCAVATION

Berger put out a call for very slim anthropologists to do the excavation and to his surprise received an abundance of interested responses. He interviewed the candidates using Skype video communication and after selecting his choices, ended up with a team of six slim, young, and science-minded women. He also had the professional cave explorers on hand to train and protect the scientists during their descent into the cave and ascent out.

The three-week excavation was funded by *National Geographic*, and it surely will be followed by many more in the future. It yielded more bones than ever envisioned (e.g., 1550 pieces), including jaw segments and a skull fragment large enough to designate the genus "Homo" as opposed to "Australopithecus" to the fossil man. The jaw and teeth also suggest a "Homo" classification. In contrast, some of the features are quite primitive.

Curved fingers of the hands indicate that the hominins had retained tree-climbing ability.

Another outstanding aspect of this find is the great number of individuals represented by these bones. There are at least 15 individuals—male, female, young and old. To anthropologists, this cross-section of an ancient population is what one hopes for but never expects to find.

11.5   DESCRIPTION OF *HOMO NALEDI*

This species of hominin displays both *Homo* and *Australopithecus* characteristics as well as traits unknown in other hominin species. Adult males stood five feet tall and females a little less. They would have weighed about 100 pounds. They were fully bipedal according to their skeletal features. The pelvis has the characteristic flaring of *Australopithecines*, but the leg, feet and ankles are those of the Homo genus (Figure 11.1).

The fingers are quite curved, more so than most *Australopithecines*, although they are better proportioned for manipulation than are *Australopithecines*. The shoulders are typically like *Australopithecines*. The vertebrae are similar to those of Homo, but the ribcage is like Lucy's species,

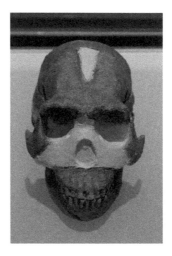

FIGURE 11.1   *Homo naledi* skull (Courtesy of Wikipedia Commons by Martinvl (own work) [CC BY-SA 4.0 (http://creativecommons.org/licenses/by-sa/4.0)], via Wikimedia Commons).

Australopithecus afarensis. Four skulls were found, and the brain volume was measured. The males averaged 560 cc. volume and the females, 465 cc. This is markedly smaller than found for *Homo erectus* (i.e., about 900 cu.cm.). The skull had features more like Homo than like Australopithecus.

Table 11.1 is a compilation of features into either homo or Australopith like groupings.

11.6 THERE WILL BE MORE TO COME

Not only is the research on the 15 fossil individuals ongoing, but eventually the Rising Star Cave will be revisited, and more fossils will be acquired and studied. When Lee Berger was asked how many bones remain in the cave, he said that he didn't know but thought there might be quite a few based on the bone harvest of 1500 pieces to date. Species like the Denisovans have been defined by a single finger joint and recoverable ancient DNA. However, in the case of *Homo naledi*, the species will be defined by an astounding quantity of fossil evidence. The opportunity to learn a lot about an important extinct hominid is exciting. There are fossil males and females of all ages, which is an extraordinary opportunity to define the species in a thorough and fact-based manner. I look forward to learning the results.

11.7 SUMMARY

TABLE 11.1 Classification of *Homo naledi* Features

Feature	Homo	Australopith
Curved toes and fingers		X
Flared pelvis		X
Precision grip	X	
Legs, knees and feet	X	
Skull shape	X	
Vertebrae	X	
Ribcage		X
Teeth	X	

A compelling and adventurous story of cave explorers finding fossil bones, recruitment of young, slim, female anthropologists to retrieve them, and the eventual retrieval and examination of those fossils makes this discovery exciting and fun. However, the find itself is spectacular. Finding 1500 hominin bones is like finding a rich gold mine to a paleoanthropologist. The hominin is most like a known species called *Australopithecus africanus*, who also once lived in South Africa. Yet, Berger sees enough human traits to assign the genus to the Homo lineage. Four skulls were found, and the brain volume was measured. The brain size of the males averaged 560 cc, while the females averaged 465 cc. These values are a little smaller than we normally consider for Homo species. However, the skulls do have features more like Homo than like Australopithecus.

CHAPTER 12

HOMO ERECTUS

CONTENTS

12.1 SCOPE

This chapter provides a description of a very important ancestor called *Homo erectus*. Fossil evidence shows that this species lived as early as 1.9 million years ago and may have still existed up to 70,000 years ago. That is an incredibly long span of time. They most likely originated in Africa, but left evidence of living in Europe and Asia as well. They were more capable of adapting to a variety of different climates than any hominin species before them. Their significance as an important ancestor comes from their astounding increase in brain size, stone tool making ability, taming of fire and cooked foods, and pioneering of the hunter-gatherer life style. They made the huge leap between man-ape and recognizable human.

12.2   *HOMO ERECTUS* IS AN IMPORTANT SPECIES

Homo erectus may be the most important to us of all the fossil men, which we discuss here. For one thing, no one seems to doubt that he is our ancient ancestor. Aside from the smaller brain, the eyebrow ridge, and the lack of a jutting chin, he is pretty much identical to us. The flaring rib cage of the australopithecines has become a vertical ribcage because *Homo erectus* has evolved a smaller gut, smaller because a diet of meat and other cooked foods had reduced the need for a large digestive tract. His new diet delivered more energy than a strict diet of raw vegetables. He used this extra energy to grow and maintain a larger and larger brain.

Secondly, *Homo erectus* is the first of our potential ancestors to make a clean break with the trees. Perhaps it was his taming of fire that made it possible for him to sleep on the ground without fear of nocturnal predators. We believe he was virtually hairless like us. He mastered the art of stone toolmaking and probably used speech of at least a rudimentary level.

Thirdly, *Homo erectus* was more capable of surviving in non-tropical climates than any of our potential ancestors of about two million years ago. We believe he evolved in Africa based upon existing fossil evidence but traveled into other continents. His fossils have been found in Europe and Asia.

12.3   THE PLACE OF *HOMO ERECTUS* IN THE HOMININ ANCESTRY

It is commonly stated that *Homo erectus* evolved from the more primitive species *Homo habilis* (2.4 to 1.6 mya), also known as Handy Man due to the stone tools associated with him. The first specimens of *Homo habilis* were discovered by Louis and Mary Leakey at Olduvai Gorge in Tanzania. They believed this species was the maker of the thousands of stone tools found in the vicinity. Unfortunately, the species *Homo habilis* has come to be regarded as a catch-all category. Moreover, there are few suitable fossils from which to designate a species prototype. A competitor for the direct ancestor of *Homo erectus* is the species *Homo rudolfensis*. One good fossil of this species was found by Richard Leakey by Lake Turkana,

Kenya. The fossil individual from this latter species was taller and had a rounder skull, flatter face, and a slightly bigger brain than *Homo habilis* (750 cc versus 640 cc) and therefore seems more like *Homo erectus* and may be the more likely ancestral species.

The term *Homo erectus* itself may include more than one species. In other words, there are differences between some of the fossils identified as *Homo erectus* sufficient to separate them into separate species. Some anthropologists restrict the term *Homo erectus* for the Asian fossils and use the term *Homo ergaster* for the African species. The Asian fossils exhibit thicker skulls and more exaggerated brow ridges and mugginess than the African fossils. In order to keep the book as simple as possible, I will use the term *Homo erectus* in the larger sense, where it includes both African and Eurasian fossils.

Some anthropologists, but not all, believe that the Asian *Homo erectus* evolved from the African *Homo erectus* and then went extinct. However, another branch of African *Homo erectus* evolved into the species *Homo Heidelbergensis*. Then one branch of *Homo Heidelbergensis* led to *Homo sapiens* while another branch led to the Neanderthals.

12.4   *HOMO ERECTUS* TRAVELED FAR

12.4.1   *PITHECANTHROPUS ERECTUS*

Table 12.1 shows several of the *Homo erectus* fossil finds. Eugene Dubois (1858–1940) was the first person to seek out human ancestral fossils and succeed in doing it. In 1887, the anatomist Dubois sailed for the Dutch East Indies with that express interest. The numerous caves in these tropical islands presented opportunities to find such fossils. He searched the caves of Sumatra, but only found fossils of living species. He then turned to the caves of Java. Near a village called Trinil, he found a skullcap and femur of a hominin. These fossils are thought to be nearly one million years old. The cranial capacity of this individual has been estimated to be 940 cc. He named the species *Pithecanthropus erectus*, meaning upright ape-man, but the species name was later changed to *Homo erectus*. When additional fossils of the same species were discovered in China, scientists of the day believed that man first evolved in Asia. However, this activity occurred

TABLE 12.1   Some of the Homo erectus Fossil Discoveries

Year of Discovery	Scientist	Fossil Location	Description of Fossils
1891	Eugene Dubois	Trinil, East Java	Skullcap and femur, 0.9 to 1.0 mya
1921	Johan G. Andersson Franz Weidenreich	Zhoukoudian, China	200 fossils of 40 individuals, including five skullcaps
1933	Gustav von Koenigswald	Ngandong, Java	Cranial capacity of 1013–1251 cc., dated at 550,000–143,000 years
1949	John T. Robinson and Robert Broom	Swartkrans, South Africa	Jaw fragment
1961	Yves Coppen	North Africa	Wind-eroded skull
1984	Richard Leakey Discovered by Kamoya Keimeu	Lake Turkana, Kenya	Turkana Boy, a.k.a. Fossil KNM-WT 15000, dated at 1.5 to 1.6 mya, is nearly a complete skeleton.
1991, 2005	David Lordkipanidze	Dmanisi, Georgia	Five skulls, one of them very complete. Brain volumes of 546 to 600 cu.cm. Dated at 1.8 mya

in the late nineteenth century, and there were no early human fossils from Africa at the time. Today, the prevailing view is that man first evolved in Africa. Moreover, *Homo erectus* is thought to have first evolved in Africa. The African *Homo erectus* is somewhat different from the Asian one and some prefer to call it *Homo ergaster* instead.

12.4.2 TURKANA BOY

The discovery and analysis of Turkana Boy (now called Nariokotome boy) in Kenya by Richard Leakey's team was significant in obtaining much of the skeleton and revealing what a complete *Homo erectus* individual is like. Still another name for this famous fossil is KNM-WT-15000. Turkana boy is easier for me to say, so I'll use that name here. The team member who first found the fossil protruding from the hillside was Kamoya Kimeu, who has other important hominin finds to his credit. Although the fossil is richly unique in the quantity of extra-cranial bones, it was no easy matter to reassemble the skull itself. Alan Walker and Meave Leakey painstakingly assembled it from numerous fragments.

A lot of research went into determining an age for this young man with estimates ranging from 7 to 18 years old. The maturation rate for modern humans is quite different than it is for chimps. The rate for *Homo erectus* is intermediate. Walker and Leakey estimated his age at 11–12 years old. Christopher Dean estimated his age at 8 years old when he died. There was also a lot of controversy regarding the dating of the fossil. Potassium-argon dating yielded a date of 2.5 million years, but that date clashed with the age of the fossil animals found in the same vicinity that were known to be much younger than that. The age of Turkana boy cited these days is 1.5 to 1.6 million years old.

12.4.3   DMANISI

Finally, Europe is added to the vast range of the *Homo erectus* migration with the discovery of a more primitive member of the species in Dmanisi, Georgia. This recent find of five primitive Dmanisi skulls, which are dated at 1.8 mya, are making paleoanthropologists reconsider their current theories. Instead of several distinct species of hominins of the Homo genus, there may have been a single Homo lineage. It seems incredible that early *Homo erectus* was able to travel so far from his African origins. The advanced tool kit (i.e., Acheulean culture) didn't appear until 1.5 mya and major brain development was still to occur at this early time in the species lifespan.

12.5   DESCRIPTION OF THE *HOMO ERECTUS* SPECIES

The most recent members of the *Homo erectus* species were not that different from modern man. They were about the same height and had a similar body. The most obvious difference would be their heads. The typical *Homo erectus* head had a smaller braincase and a bony ridge over the brow, and lacked a chin. Figure 12.1 is a typical skull. The individual may have been hairless except for the head and pubic area like modern man. Figure 12.2 is a reconstruction of the head. Scientists believe that he was the first hominin to migrate out of Africa. *Homo erectus* had come a long way from the previous bipedal apes in his ability to live where there were no suitable trees to climb for safety. He had adapted to living and sleeping on the ground in a variety of terrains. He developed the use of fire and cooking. The astounding increase in brain size tells us that having superior intelligence was of paramount importance to his survival and/or procreation. He had become a competent hunter-gatherer over time. He may have been a meat scavenger rather than a hunter for many generations, but his stone toolmaking skills and stalking prowess had advanced to where he could obtain meat on a regular basis via hunting of various sized game.

12.5.1   STONE HAND AXES

Hundreds of stone hand axes have been found in sites associated with *Homo erectus*. The typical hand axe had a roundish shape, except it was

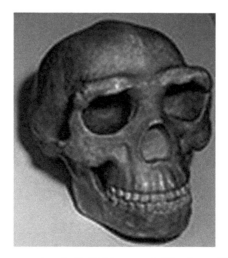

FIGURE 12.1 *Homo erectus* skull (Peking man) (Courtesy of Wikimedia Commons by Kevinzim [CC BY 2.0 (http://creativecommons.org/licenses/by/2.0)], via Wikimedia Commons).

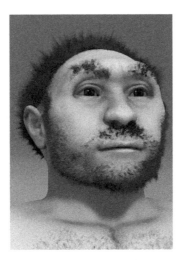

FIGURE 12.2 Reconstruction of *Homo erectus* (Courtesy of Wikimedia Commons by Cicero Moraes (own work) [CC BY-SA 3.0 (http://creativecommons.org/licenses/by-sa/3.0)], via Wikimedia Commons).

tapered and pointed at one end. They were chipped around the entire periphery to produce a sharp edge. Although they were somehow important to *Homo erectus* as indicated by the large numbers found, it is hard to see what they were specifically used for. Neurobiologist and prolific author William Calvin suggests they were used in hunting by throwing them at prey. Marek Kohn and Steven Mithen independently arrived at the idea that they were important in sexual selection. Perhaps women selected men who had the skill to fashion a symmetrical and shapely hand axe as mates.

12.6 VANISHED CARNIVORES

Lars Werdelin has a hypothesis, that, if true, gives us an insight to *Homo erectus* and his ancestors of nearly two million years ago. Werdelin is an expert on the evolutionary history of predators in the savannahs of Africa and he sees a big anomaly. He wrote an article for the *Scientific American* magazine (November 2013 issue), "King of the Beasts," where he presents his ideas on human effects on carnivore populations.

The number and diversity of carnivores took a nosedive about 1.5 million years ago. He conjectures that meat-eating *Homo erectus* might be responsible for the disappearance of many carnivore species. By 300,000 years ago, the winnowing was complete, and the surviving carnivores resemble those seen today. However, the carnivores of today are mainly hyper-carnivores, who eat meat for 99% of their diet. Gone is the functional richness that carnivores once had where many species existed that had mixed diets of meat and vegetation.

Werdelin cites anthropologist Henry Bunn, who sees the Homo lineage diet transitioning in three steps: Those steps are presented in Table 12.2.

There is a dearth of homo lineage fossils for the first two stages, but *Homo erectus* fossils and his characteristic stone tools exist for the third stage. Yet, the timing of meat-eating man coincides with the disappearance of carnivore species. At the end of the article, Werdelin reminds us that the introduction of a new predator can have significant effects on an environment. For example, the reintroduction of wolves into Yellowstone Park markedly transformed the area. The wolves trimmed the elk herds, substantially resulting in thriving plant growth. This in turn fostered the reappearance of beavers and other animals. So it is quite possible that *Homo erectus* significantly altered the functional richness of carnivores in Africa during his transition to a hunter-gatherer life style.

12.7   SUMMARY

Homo erectus is a very important species for several reasons. He was obviously our ancestor based on how similar he is to us. Except for the smaller brain and heavy eyebrow ridges, he could probably pass as one of us. This species is also important to our evolutionary history because it underwent huge anatomical adaptations as it became proficient in a hunter-gatherer

TABLE 12.2   Stages in Homo Lineage's Meat-Eating Transition

Time Period	Stages in the Transition to Meat-Eater
2.6–2.5 mya	Occasionally butchering on bones with stone tools
2.3–1.9 mya	Transporting meat-rich carcasses and splitting bones to retrieve marrow
1.8–1.6 mya	Extensive butchering and likely hunting of animals

life style. This species evolved a three times bigger brain than apes have, evolved longer legs for running, evolved shorter arms than apes have, lost its thick fur in order to enhance cooling via sweating, and became an endurance runner. *Homo erectus* was also the first hominin species to migrate out of Africa into Asia and Europe. As a hunter-gatherer, he became very skilled at making stone tools and weapons, tamed fire, developed cooking, and probably communicated via speech.

CHAPTER 13

NEANDERTHAL MAN

CONTENTS

13.1   SCOPE

This chapter is about a fascinating species of human, the Neanderthals. Although they are a different species than we are, there is evidence that interbreeding between us occurred. Fossils of this now extinct species were popping up in Europe and the Middle East area even before Darwin wrote *Origin of the Species*. We know a lot about Neanderthals due to the numerous fossil finds. This means that we can draw conclusions about them from data instead of merely speculating.

13.2   GENERAL

The official name of the species is *Homo neanderthalensis,* but I am going to just call them Neanderthals. Their fossils were the first hominin species to be recognized as different from us. These fossils were more readily discovered because these people only recently went extinct (i.e., 30,000 or so years ago). Recent fossils are easier to find than more ancient ones because there has been less time for damaging geological events to destroy them. That explains why Neanderthal fossils have been found at numerous sites in Europe and the Middle East. Aside from our own *Homo sapien* fossils, the most frequently found fossils are those of the Neanderthals. They are essentially contemporaries of ours, only recently going extinct after coexisting with our ancestors in Eurasia for many generations.

The Neanderthals were physically adapted to cold climates in ways that our species is not. For example, their nasal system was larger and different from ours. They were highly dependent on a meat diet and they acquired that meat using a hunting style known as "ambush hunting." What that means is that they lay in wait for their quarry and then surprised and attacked them at close range with thrusting spears. We can see from their bones that they had shorter and squatter bodies than our Cro-Magnon ancestors and were probably much stronger. Compared to the *Homo sapiens* of 40,000 years ago, Neanderthals had thicker bones, bigger and stronger muscles, and had closer encounters with the big game that they hunted. Our ancestors, by comparison, were taller, lighter-boned and muscled, and used projectile weapons to hunt dangerous game from a safer distance. Neanderthals had brains equal or bigger than those of *Homo sapiens*, although the shape of the skulls differed. See Figure 13.1. Our frontal lobe area is larger than for a Neanderthal, giving our skulls a more roundish appearance. If you put a business suit on a Neanderthal man and had him ride the subway, there would be a lot of people staring at him. He had a huge broad nose, a protruding eyebrow ridge, and lacked a chin. Neanderthals were not as stupid or as primitive as some people believe. Consider that they had survived in a very harsh climate for over 100,000 years, made stone tools and weapons, and used fire to keep themselves warm. Their main disadvantage in competing with the Cro-Magnon newcomers was that they lacked our spirit of using innovation to solve problems.

13.3   FOSSIL FINDS

Neanderthals lived in Europe and parts of Western Asia during the Ice ages. Recent fossil evidence has extended their range to as fareast as Siberia. There is also evidence that early Neanderthals and early *Homo sapiens* occupied the same sites in Israel around 100,000 years ago, although not

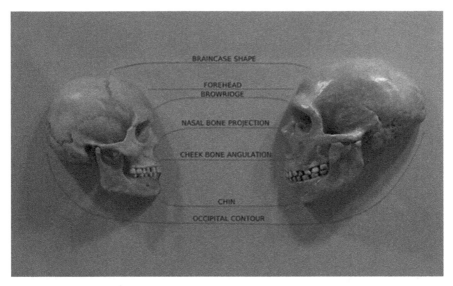

FIGURE 13.1 Comparison of human (left) and Neanderthal (right) skulls (Courtesy of Wikimedia Commons by hairymuseummatt (original photo), Kater Begemot (derivative work) [CC BY-SA 2.0 (http://creativecommons.org/licenses/by-sa/2.0)], via Wikimedia Commons).

necessarily at the same time. Before going extinct, Neanderthals made their final bid for survival in the Gibraltar area, occupying a network of caves. It was there that they finally went extinct. Table 13.1 shows some of the Neanderthal fossil discoveries.

13.4 DIFFERENCES BETWEEN HUMANS AND NEANDERTHALS

The examinations of teeth from Neanderthal fossils suggest that Neanderthals matured faster than humans. For example, the growth lines in an eight-year-old Neanderthal are similar to the growth lines in a twelve-year-old human.

The science of DNA will be introduced later in the book, but it is helpful to know now that human DNA is an extremely long double-stranded molecule. The two strands are connected to one another by base pairs, which carry a four-letter code. DNA from two individuals or two different species can be compared by matching up the sequence of base pairs and

TABLE 13.1 Some of the Neanderthal Fossil Discoveries

Year Found	Scientist	Fossil Location	Age of fossil	Description of Fossils
1829	Philippe-Charles Schmerling	Awirs Cave, Belgium	30–50K	Engis 2, Cranium, fragments of upper and lower jaw
1848	Captain Edmond Flint	Forbes Quarry, Gibraltar	30–40K	Gibraltar 1, nearly complete skull
1856	Johann Fuhlrott	Feldhof cave, Germany	40K	Piece of the skull and large bones
1908	A & J. Bouyssonie and L. Bardon	France	60K	La Chapelle-aux-Saints 1, partial skeleton
1926	Dorothy Garrod	Devils Tower, Gibraltar	Undated	Gibraltar 2, fragmented skull of a child
1938	R. Capitan and D. Peyrony	Uzbekistan	70K	Teshik-Tash Skull, Child's skull assembled from 150 bone fragments
unknown	Dorothy Garrod?	Tabun Cave, Israel	120K	Near-complete female skeleton
1957	Ralph Solecki	Bradost Mountains, Iraq	46–60K	Shandikar 1 thru 4 1: elderly deformed male skeleton, 2: Crushed male skull and bones, 3: Male with stab wound, 4: Male in fetal position
2001	Luc			
2007	Anatoly Derevianko	Southern Siberia		Small bones, confirmed by DNA analysis
Unknown	Finlayson	Rock of Gibraltar	24–28K	Mousterian tools, campfires ash, animal bones, etc.

Source: https://en.wikipedia.org/wiki/List_of_Neanderthal_fossils.

identifying diffcrences. Scientists have developed procedures for restoring and examining the DNA of ancient individuals. Consequently, we
now can compare Neanderthal DNA with that of modern humans. Here is
what we have learned.

Neanderthals were not that different from our species, *Homo sapiens*,
differing in our DNA by only 0.12 percent. This means that 99.7% of the
base pairs were identical. They had a large brain (1600 cu.cm.) to go with
a big body. Humans have a cranial capacity of about 1400 cu.cm by comparison. Neanderthals were adapted to living in a cold climate. They had
a very large nasal area compared with our species or any other hominin
species. Ian Tattersall found a quite unique feature in the nasal area of
Neanderthal skulls. Inside the rim of the huge nasal opening, he noticed
that there are large bony projections on each side. Behind those projections lies an even larger swelling on the inside of the nasal cavity. He has
seen this feature on every Neanderthal skull that was well-preserved. He
reasons that this unique feature must have had some important benefit to
their breathing ability. Only Neanderthals have this nasal feature. It is not
seen in any of the other species in the Homo lineage except them.

13.4.1 NEANDERTHAL ANCESTRY

Neanderthals are thought to be descended from Heidelberg man (*Homo
heidelbergensis)*, a species that lived between 600,000 and 200,000 years
ago. Heidelberg man, in turn, is thought to have descended from *Homo
erectus*. Heidelberg man fossils have been found in Africa, Europe, and
western Asia. The species name came from a mandible found near Heidelberg, Germany, in 1907. However, the most scientifically valuable fossil
find was in a cave in Atapuerca, Spain, in 1997. More than 5,500 bones
were found dated at 350,000 years old. Thirty-two individuals are represented by the bones. They are so similar in features to Neanderthals that
some researchers think they should be designated as Neanderthals. Heidelberg man fossils were also found in northern Rhodesia in 1921. The
"Broken Hill Skull" is the most famous of them.

A lot is known about Heidelberg man. The male's height averaged 5 ft.
9 in. and the females averaged 5 ft. 2 in. Their cranial capacity averaged
1100 to 1400 cm^3, like modern humans but smaller than Neanderthals.

They were predominately right-handed, buried their dead, and could probably talk. It is also thought that Heidelberg man was the species that migrated out of Africa into Europe and evolved into Neanderthal Man. The physical changes came from adapting to a colder climate and adapting to ambush hunting, which required great strength and robustness.

13.4.2   NEANDERTHAL TOOLS AND WEAPONS

Dated at about 400,000 years ago, the butchered remains of 10 horses were found in a coal mine at Schoeningen. Three wooden spears several meters long were also found there, which seem to be designed for throwing. They were made from the trunks of spruce trees. They were sharpened at the end, which was the base of the tree and where the wood is hardest. The thickest and heaviest part of the carved shaft was one-third of the length back from the projectile end.

Neanderthals were ambush hunters of large animals. These close encounters with powerful animals frequently resulted in injuries to the hunters. Examinations of Neanderthal bones show that they sustained injuries much like those of rodeo riders. They used wooden spears with stone points to kill their prey. Now let's consider their stone toolmaking abilities.

The earliest stone tools were found at the Le Moustier site in France. The Neanderthals continued making stone tools in this style for one hundred thousand years. That style is called "Mousterian" in honor of the site. The toolmaking process began with the shaping of a stone core. A single blow could then detach a sharp flake. Flint was commonly used for stone toolmaking. Spear points, axes, and knives were common stone weapons. Scrapers were also used to scrape meat from bone.

13.4.3   INTERBREEDING

Later in this book, an entire chapter (Chapter 28) is devoted to the topic of Neanderthal-Human interbreeding. We now know that this interbreeding actually happened due to evidence in our genetic code. However, we never would have known were it not for the persistence and ingenuity of Svante Pääbo. He made it his life's work to resurrect the DNA of long extinct beings, both animals and man. Initially, he focused on mitochondrial DNA (mtDNA) rather than nuclear DNA of long dead beings. Mitochondrial DNA is more easily obtained and is a much shorter molecule. When Pääbo was able to sequence the mtDNA of Neanderthal bones and compare it with human mtDNA, it looked doubtful that any interbreeding had occurred.

It was years later after conquering a series of technical difficulties that Pääbo was able to sequence Neanderthal nuclear DNA. We are talking about a 3.2-billion nucleotides-long molecule versus a mere 16,500 nucleotides for mtDNA. (Note: a nucleotide is the repeating organic unit in DNA.) The DNA results were quite different. Many humans do carry Neanderthal contributions to their nuclear DNA, which amount to about two percent. This interbreeding evidence was not detectable in the mtDNA results.

13.4.4 NEANDERTHAL EXTINCTION

13.4.4.1 Adaptation to Ambush Hunting

Fifty thousand years ago, Eurasia was populated by the Neanderthal people. However, somewhere between 30,000 and 40,000 years ago, they had all disappeared. Some anthropologists suspect that our ancestors may have killed off the Neanderthals, but Clive Finlayson has studied the Neanderthal culture extensively and is skeptical of those theories. Instead, he believes that climate change had doomed them already. He felt that over tens of thousands of years of ambush hunting, the Neanderthal's body had become over-specialized to endure close encounters with powerful prey. They had become thick-boned and highly muscular, which was advantageous in ambush hunting. However, when colder climates reduced the woodlands to treeless areas, ambush hunting was no longer feasible. This climate change was a cooling trend that began forty-four thousand years ago and reached a low point thirty-seven thousand years ago.

13.4.4.2 The Last Refuge

Finlayson also tells about the last refuge for the Neanderthals before they disappeared. It was in the caves of the Rock of Gibraltar. Located on the southern coast of Spain, Gibraltar has the Mediterranean Sea to the east and the Atlantic Ocean to the west. The continent of Africa seems to be a stone's throw across the narrow Strait of Gibraltar to Morocco when viewing a map. It was here mainly in Gorham's Cave, but also in other nearby caves that a wealth of evidence helps paint the picture of the Neanderthal's final period. Amongst that evidence is charcoal from many fires, pollen, bones from the animals they ate, and their stone tools. The infill at the bottom of the Gorham Cave is 18 meters deep. The lowest level dates to 125,000 years ago. The Neanderthals lived here for 4000 years (28,000 to 24,000 years ago).

Finlayson did extensive fieldwork to establish the type of climate that existed during the Neanderthal's time here. He concluded that the climate was temperate and dry. This is in contrast with the rest of Europe, where a cold treeless landscape made ambush hunting impossible. Here in Gibraltar, there was a mosaic of shifting sand dunes, wooded savannah, and thickly wooded areas near streams. Seasonal lakes attracted ducks and other waterfowl. Herds of grazing animals were hunted in the area. They included wild boar, aurochs, red deer, horses, narrow-nosed rhinos, and straight-tusked elephants. Moreover, ibex were taken higher on the Rock of Gibraltar itself. Rabbits must have been easiest to catch because they were eighty percent of the Neanderthal diet. Yet, evidence of flint knife-nicked animal bones charred by the campfires shows they took larger animals as well.

We do not know what killed off these last of the Neanderthals, although it is known that small populations are especially in danger of extinction. Perhaps a disease took the last of them out.

13.5   SUMMARY

Neanderthals, formally known as *Homo Neanderthalensis*, is a sister human species to us that was adapted to the cold Ice Age climate of Eurasia. They were a large-brained, hominin distinguished from us by a low forehead, prominent eyebrow ridge, large nose, and absence of chin. Moreover, they were bigger boned and muscled than us. Whereas our Cro-Magnon ancestor in Europe hunted at a distance using projectile weapons, Neanderthals were ambush hunters, which caused them to be sometimes injured from their close contact with large prey animals, like horses, aurochs, deer, rhinos, and others.

The Neanderthals were most likely descended from *Homo Heidelbergensis*, a species descended form *Homo erectus*. Neanderthals have existed in Europe form 120,000 years or longer. They went extinct shortly after our ancestors arrived in Europe around 40,000 years ago. Their last occupation site was in Gibraltar, Spain, where the climate was warmer than northern Europe.

Genetically, Neanderthals are very similar to *Homo sapiens*, even if our appearances are so different. It is now known that we interbred with them and we currently carry about 2% Neanderthal DNA in our genomes.

CHAPTER 14

HOMO SAPIENS

CONTENTS

14.1   SCOPE

We began our journey by coming *down from the trees*. We began with our common ancestor with today's chimpanzees of some 5–7 million years ago and have progressed through a trail of fossil men that gradually looked and acted more like us. Finally, that trail leads to our own species, *Homo sapiens*. We are the last man standing; the only hominin species not to go extinct. We are also unique. Anyone can spot a *Homo sapiens* skull from a lineup of hominin skulls. Our skull looks quite different; it is more rounded, lacks an eyebrow ridge, has a flatter face, smaller mandible, and a jutting chin.

Our species has been around for about 300,000 years and hasn't changed much in skeletal appearance during that time, but huge changes occurred culturally. We have transitioned from a hominin, living a repetitious primitive existence with little innovation, to one showing profound innovation in everything we do. That ability to innovate, communicate, and build on previous discoveries led to a species that flies in airplanes, projects images across the world in seconds, walks on the moon, cures diseases, and regards almost nothing as impossible.

14.2   THE AGE OF OUR SPECIES

14.2.1   GENETIC EVIDENCE

Until 1987, nearly everything we knew about our prehistoric ancestors came from fossil evidence, ancient campfire residue, stone tools, and tool marks on animal bones. Then in 1987, three authors presented a paper that has given us powerful new tools to explore our prehistoric past. Rebecca Caan, Mark Stoneking, and Alan Wilson published a paper in *Nature,* which rocked the scientific world. Using DNA analysis techniques, they determined the age of the most recent common female ancestor of all living people. Essentially, they traced the line from mothers to daughters back progressively in time. This woman was nicknamed "Mitochondrial Eve." She was estimated to have lived 200,000 years ago. The details of these analyses are presented in Chapter 27.

These studies also established that *Homo sapiens* originated in Africa. The limited diversity in the DNA of living humans around the globe suggests that the human population size went through a bottleneck at some previous time. The mutational markers showed that a migration out of Africa occurred about 50,000 years ago. The dominant view of how our species managed to populate the Earth is called the "Out of Africa" theory. However, some anthropologists still strongly adhere to the opposing theory, the "Multiregional theory," where it is thought that modern humans evolved from early *Homo erectus* in distinct parts of the world. This is discussed in more detail in Chapter 28. The pure interpretation of this theory would have this human territorial expansion account for all the human population today. The newer, modified, Out of Africa theory has modern man cross-breeding with *Neanderthals*, *Homo erectus,* and other archaic humans. This topic is explored in more detail in Chapter 30.

14.2.2 OUR OLDEST AFRICAN ANCESTORS

Recent fossil evidence confirms a 300,000-years-old age for our species. However, before that discovery, Tim White had led an excavation that found two adults and one child in the Afar region of eastern Ethiopia. The fossils have been dated at 160,000 years ago. The team also unearthed skull pieces and teeth from seven other hominid individuals, hippo bones with cut marks on them, and over 600 stone tools including hand axes. Finding *Homo sapiens* living in Africa this long ago proves that we did not descend from Neanderthals, as some had suggested. Figure 14.1 shows the skull of that ancient human.

Another ancient find for our species was conducted by a team directed by Richard Leakey between 1967 and 1974. The hominin bones were found at the Omo Kibish sites near the Omo River in Ethiopia. The site has been dated at 190,000–200,000 years old. Leakey has determined that some bones are *Homo sapiens* and they would then be the oldest bones ever found for our species.

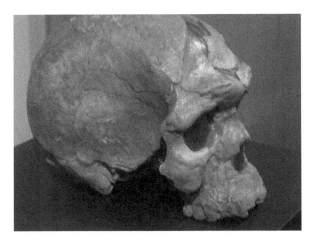

FIGURE 14.1   Tim White's ancient human (*Homo sapiens idaltu*) (By Alessandrosmerilli (Addis Ababa national museum) [Public domain], via Wikimedia Commons).

14.3   THE CAMPANIAN EXPLOSION HELPS DATE PREHISTORIC SITES

The initial population of our ancestors in Europe and the Near East was only a few thousand people, so evidence of their presence was usually scant. In terms of fossils found, it was sometimes as little as a single human tooth or a tool-marked animal bone. Stone chips and fragments from stone tool making was another common indication that our ancestors were at a certain place.

Well, something horrific happened 39,000 years ago. It was horrible for the people and animals in a wide radius around what is now Naples, Italy. It was a huge volcanic eruption; so huge that it left a caldera with an area of 89 square miles. It had a huge ash plume that rose 27 miles and then deposited that ash over the eastern Mediterranean and 1550 miles northward across the Balkans and Eastern Europe. It is called the Campanian event. What was so horrific for the victims was a blessing for the archeologists. These widespread ash deposits provide a reference stratum for dating the skimpy number of *Homo sapiens* sites. There had been *Homo sapien* migrations in the ash zone prior to the Campanian event, but the evidence after the event demonstrates a widespread linkage and a common unique stone tool culture. That culture is the Aurignacian culture.

14.4   THE AURIGNACIANS

The Aurignacians of the upper Paleolithic are known by the unique stone and bone tools, which they made. They lived at a time (45,000 to 35,000 years ago) when the climate was cooling and woodlands were giving way to treeless areas. These hunters no longer had the cover of trees but had to pursue game in areas where they could be seen approaching them in the open. A new type of weapon was needed; one that could be launched from a distance. This meant lightweight spears propelled with a spear-thrower.

Aurignacian spear points were often fashioned from bone or antler with grooves cut in the bottom. Flint tools include fine blades struck from prepared cores. Due to the sophistication involved in the Aurignacian toolmaking, archaeologists conclude that they were modern humans. However, the fossil evidence for that conclusion is sparse.

Figurines of now extinct mammals have also been found at Aurignacian sites. These include mammoths, rhinos, and Eurasian wild horses (i.e., tarpan). Bone flutes have also been found at these sites. However, the greatest surprise about the skill and imagination of the Aurignacians occurred when the cave at Grotte de Chauvet was discovered. The art revealed on the walls of this cave was astonishing. The oldest paintings have been dated to 36,000 years ago, although some may be as young as 24,000 years ago. For example, on one wall, nine lions and a single reindeer face to the left, while seventeen rhinoceroses cavort on the right. In another area, there is an owl with his head turned 180 degrees. The paintings go on and on, with a level of artistic skill that is outstanding. Brian Fagan in his book *Cro-Magnon* concludes that the Aurignacians were responsible for many of the paintings in this cave.

14.5 CRO-MAGNON MAN

Cro-Magnon people were *Homo sapiens*, but not all living *Homo sapiens* are descended from the Cro-Magnon. They were a group of *Homo sapiens* that migrated into Europe around 35,000 years ago and left evidence of themselves. In fact, the site that bears their name was near Les Eyzies in France. The Cro-Magnon fossil site, discovered in 1868, was dated at 30,000 years ago, and contained skeletons of three adult males, an adult female, and a child. The Cro-Magnon people were nearly identical to people existing today, were hunter-gatherers adapting to an Ice Age environment, and were more innovative than their ancestors as demonstrated by advanced stone-and-antler tools, skillfully executed cave art, figurines, and bone flutes.

14.6 COMPARISONS OF CRO-MAGNON MAN WITH NEANDERTHALS

The Cro-Magnons are often compared to the Neanderthals, who co-existed with them for thousands of years but who finally went extinct. Why did one group survive while the other group died off? The environmental conditions of the Ice Age, when this drama took place, were variable and severe. Yet, the Neanderthals had physically adapted to the freezing conditions over tens of thousands of years. The Cro-Magnons were relative newcomers to Eurasia, having tall, slender bodies adapted to the warmer African savannahs. It would seem that the Neanderthals had the advantage, but the opposite was true.

One of the reasons for the Cro-Magnon survival advantage was that they used sewing to fashion warm clothing, which was layered and with sealed-off air gaps. Cro-Magnons could fashion such clothing because they had invented the eyed needle, made of antler material and drilled through using flint microchips. It is believed that the Neanderthals had no ability to sew and wore animal pelts, which fit loosely about them. It is even thought that the Neanderthals lacked footwear and walked barefoot on snow and ice. It is not unimaginable. Charles Darwin reported seeing naked Indians adapted to the cold climate of Tierra del Fuego on the southern tip of South America, so we know it is possible to make this kind of adaptation.

14.7   THE GRAVETTIANS

The Gravettians lived in a harsh, cold, treeless environment and yet made huge strides on the path to becoming modern humans. The term "Gravettian" is used by archeologists to designate a distinct stone toolmaking culture. The Gravettian style is different enough from the Aurignacian style that we can see the evolution in toolmaking. A lot is known about the people associated with those tools. The name comes from the La Gravette site in the Dordogne region of France, where the characteristic tools were first found. It dates between 28,000 and 22,000 years ago.

The Gravettians formed the first sizable, settled villages in Europe. Until their villages, most humans of the time lived in small groups of migrant hunter-gatherers totaling 30 or less individuals. These nomads lived in tee-pee style tent structures. The Gravettians, by contrast, dug oval or circular pits in the loess and arranged mammoth bones around the pits to support hides and other coverings. They also made structures for storing meat. The temperatures of the environment were cold enough so that refrigeration was automatic. Game was plentiful and consisted of large herds of reindeer, steppe bison, horses, and other large animals.

14.7.1 SPECIALIZATION OF LABOR

Living in larger than normal groups, the Gravettians brought a new concept to humanity: the division and specialization of labor. Obtaining food had become an easier part of living due to the abundant game and the superior methods of killing and trapping it. Storage of food was also easy due to the cold climate and the storage provided nourishment during those lean hunting periods. The Gravettians actually had free time to be creative. They used that time to get better at making clothes, weapons, tools, and other crafts.

14.7.2 VENUS FIGURINES

One thing the Gravettians liked to make was the Venus figurines. We know this because they have been found in various parts of Europe, from Ukraine to the southwestern France. The Venus figurines depict naked women with large hips and breasts, sometimes showing the vulva explicitly. Although many are made of baked clay, others are carved from ivory. The clay figurines may be the earliest known ceramic. The Venus figurines may have been an expression of sexuality and what was deemed feminine beauty in Gravettian times. Explicit wall paintings show that sex was on the minds of these earlier humans 25,000 years ago.

14.8 SOLUTREANS

The Solutrean culture was both magnificent and brief. The site, which it is named for, is located at Solutre in east-central France, and the characteristic stone tools have been dated between 22,000 and 17,000 years ago. The technique involved percussion and pressure flaking rather than the cruder knapping process. Beautiful projectile points were fashioned in the shape of laurel leaves. The Solutreans were hunting reindeer and other cold-adapted animals as early as the extreme of the glacial cycle (i.e., The Last Glacial Maximum).

The Solutreans may have been skilled cave art painters as well as hunters. The cave at Lascaux, France, contains 600 paintings of animals and signs plus 1500 engravings. There are breath-taking paintings of horses, aurochs, stags, a bear, and a mythic beast and these are in the first chamber alone. It is reasoned that these polychrome paintings were done between 19,000 and 17,000 years ago based on radio-carbon-dated artifacts on the floor. Abbe Henri Breuil called Lascaux "the Sistine Chapel of Prehistory."

14.9   MAGDALENIANS

The Magdalenian culture was a long-lasting epoch beginning 17,000 years ago following in the steps of the Solutreans and ending 12,000 years ago with the disappearance of the animal herds, which they pursued. Their population increased markedly during the Bolling oscillation, a 1500-year period when the climate was much warmer, and life was more certain. The name Magdalenian is derived from the rock shelter at La Madeleine. Evidence of their presence at this site fills parts of several museums. The use of bone and ivory, which had accelerated in the Solutrean culture, was increased even more so in the Magdalenian culture. These bone implements included spear points, harpoon-heads, needles, hooks, and borers. Their most recent toolmaking trended toward microliths.

The Magdalenians were hunter-gatherers, who followed the herds of horses or reindeer on the continental tundra. Huge bone piles have been found at places where they regularly camped in their tents. In addition to the large game animals, they hunted and trapped smaller animals. They also harvested salmon when they made their annual upriver migrations to breed. Yet, the Magdalenians also were skilled artists and artisans. They contributed to the prehistoric cave art at Lascaux, but the famous cave at Altamira in Cantabria, Spain, may have been the chief meeting site for their bands. The Magdalenians used color pigments to a greater extent than the previous cultures and developed a new technique of application, namely, spray-painting. There is a long list of cave sites where the Magdalenians left their art.

Ceramic art was practiced over 15,000 years ago in Croatia. It was known as Vela Spila pottery. Moreover, they crafted items for adornment such as bracelets, pendants, necklaces, and others. Jewelry art made from ivory, bone, or antlers was common. Ivory carvings were covered with fine figurative and geometric engravings. Towards the end of their culture, the artistic quality seemed to fall off.

14.10   POPULATING THE PLANET

Homo sapiens didn't only migrate into Europe, but eventually migrated to every continent on the Earth except Antarctica. There is evidence that *Homo sapiens* migrated to Australia by 46,000 years ago. This would have been concurrent with their first arrival in Europe. North and South America had never seen a human being until northeast Asians crossed the Bering Land Bridge about 15,000 years ago. Alternatively, they may have traveled in primitive watercraft. In any case, they quickly filled both American continents. The South Pacific islands were not populated until 5000 years ago.

Genetic evidence gathered from living humans from around the world has been used to track the migrations from the time *Homo sapiens* left Africa up until the migrations ended. Chapter 30 tells this story.

14.11   OVERVIEW OF THE FOSSIL MEN

The initial question was how is it possible that we are so closely related to African apes and yet be so radically different from them. We have just reviewed the fossil evidence and it tells us part of the story. Bipedal walking preceded the evolution of bigger brains by millions of years. There was a very gradual adaptation of apelike creatures from being exclusive tree-dwellers to living in open spaces with fewer and fewer trees. This was a necessary adaptation because climate change turned tropical forests into open woodlands and then into savannahs and deserts. This was a process that occurred over millions of years. Big brain development first began about 2.5-million-years-ago with a branch of hominins, whom we designate as the genus Homo. The earliest of the genus (*Homo habilis* and *Homo rudolfensis*) were quite similar to the Australopithecines. The link between *Homo habilis* and the more recent Homo species is *Homo erectus*. This species underwent remarkable changes in both mind and body. Its fossils have been found from as early as nearly two million years ago and as late as 300,000 years ago. It has progressively changed over that time span to become more and more like modern humans. The latest of the genus homo (e.g., *Homo Heidelbergensis*, *Homo Neanderthalensis*, and others) look more like us. They had large brains; had tamed the fire; made clothing, tools, and weapons; and could adapt to nearly any climate.

One thing still missing from this scenario is how is it possible for one kind of creature to turn into another kind of creature? How do we know that these fossil men were really related to each other? It sounds magical. What is the scientific basis for evolution? That is what we will consider in the next part of the book.

14.12   SUMMARY

Our species is called *Homo sapiens*, which means "wise man." We are a unique species, easily distinguished from the other species that preceded us. Our rounded skulls, free of bony eyebrow ridges, is unique, but another feature entirely human is our jutting chin. *Homo sapiens* have been on this Earth for 300,000 years at least. We know of our prehistory both from fossil evidence and from genetic evidence. Mitochondrial DNA gathered from women around the world helped identify that common human female ancestor, Mitochondrial Eve. She lived around 200,000 years ago in Africa. Later, the male common ancestor was traced using the Y-chromosome. A similar age was obtained for Y-chromosomal Adam.

A behavioral change in our species seems to have occurred some 50,000 years ago. After many thousands of years of exhibiting little to no change in their stone tool culture, our species became hyper-innovative in what has been called the Great Leap Forward. Innovation was exhibited in not only stone toolmaking, which advanced rapidly, but in all aspects of human endeavor. Even activities which were non-essential to survival were practiced such as artistic portrayal of animals, fashioning statuettes of females and other items, playing music on bone flutes, making decorative jewelry, and ritualistic burials.

PROBLEM SET FOR PART II

QUESTIONS

Q1. Carbon dating is limited to excavation sites less than 50,000 years old. Why is that?

Q2. Which paleoanthropologist is associated with the discovery of 3.5-million-years-old hominid footprints in Africa?

Q3. The general rule is that two similar species cannot long exist together in the same area. Yet, *Homo erectus* and *Paranthropus boisei* coexisted for hundreds of thousands of years. Explain.

Q4. Some doubt that *Ardipithecus ramidus* is a true hominin ancestor due to a characteristic of their feet. What was it?

Q5. Who helped Donald Johanson analyze the fossils from the First Family and conclude they comprised a single new species?

Q6. Some anthropologists do not think *Australopithecus sediba* should be a separate species. Where do they think it belongs?

Q7. Search the Internet for "*Homo naledi*" and write a few sentences about what has recently been discovered. Identify your sources.

Q8. One of our ancestors made symmetrical stone hand axes and migrated out of Africa over a million years ago. Who was it?

Q9. Archeologists excavating a 50,000-years-old site in Spain find stone tools of the Mousterian culture. What human species do you think made these tools?

Q10. When archeologists identify a site visited by early *Homo sapiens* as Solutrean or Magdalenian, what does that mean?

PART III

HOW EVOLUTION WORKS

Our quest is to understand how it is possible that a tree-dwelling ape could over the course of a few million years be transformed into a human being. Of course, humans are not the only animals to be transformed by evolution. Whales were once quadrupedal land animals. Birds were once a feathered form of dinosaur. There are countless more examples that could be cited. Evolution is the foundation of the biological sciences.

In order to properly understand the evolutionary history of man, we are going to have to understand what evolution is and how it works. In this section, we shall see how Charles Darwin awoke the world to the reality of evolution and the forces driving evolutionary change. However, scientists in Darwin's day had much to learn about how genetic traits are passed from parents to offspring. The modern synthesis updated Darwinian evolutionary theory with that knowledge as it developed over the decades.

In our own age, a person arose to defend Darwin's views and write in a manner that educated the public. Richard Dawkins is that man and he has been a prolific author. He is also associated with linking evolution to a competition between genes. We tell his story too.

I also try to convey how environment change produces stresses that cause evolutionary change. In some cases, new species form. There are also the dynamic struggles of co-evolution to consider, for example, the ongoing struggles between predator and prey or parasite and host. These battles for survival span generations and give rise to the Red Queen Effect. One has to constantly run just to stay in place. Moreover, sexual reproduction has advantages over asexual reproduction in these struggles.

Finally, we apply our newly gained knowledge to the evolution of man and his ancestors.

CHAPTER 15

CHARLES DARWIN

CONTENTS

15.1 SCOPE

Charles Darwin (Born: 12 February 1809; Died: 19 April 1882) made the concept of evolution popular throughout the world. Most people today recognize Darwin as the father of evolutionary theory and associate the phrase "survival of the fittest" as the driving force behind evolutionary change. His 1859 book, *On the Origin of the Species by Means of Natural Selection*, changed the way the world thought about the origin of living things, ourselves included. However, fewer people realize what a remarkably great scientist Darwin was and how dedicated he was in studying and revealing the secrets of life. In this chapter, we shall discuss Darwin's important contributions. He not only proved that life forms evolve but explained how evolution works.

15.2   HIS BOOKS CONTAIN HIS LIFETIME ACCOMPLISHMENTS

Charles Darwin devoted a lifetime to investigating and reporting on the mysteries of the world. Table 15.1 compiles many of his accomplishments into a single chart. He lived in a time and in a social environment that considered that all questions had already been answered in the Bible and when challenging the Church on its positions was deemed unwelcome. Yet there were other scientists like Darwin, who were pioneers in an age of discovery. Charles Darwin was influenced by Jean Baptiste Lamarck, Thomas Malthus, Comte de Buffon, Eramus Darwin, Charles Lyall, George Cuvier, and James Hutton. It was their ideas that started Darwin down the path to his great discoveries. Science was advancing by new generations of scientists building on the shoulders of giants. Darwin has become one of those giants, perhaps the greatest of them all.

Readers, living in the 21st century, may find the writing style of Darwin's day different from what they are used to reading. People talked and wrote differently in the nineteenth century. I found that even when he was in his twenties, Darwin had a very rich vocabulary and was able to express his observations in great detail (Figure 15.1). Here, I am referring to his experiences as a naturalist on the H.M.S. Beagle. I found his observations

TABLE 15.1   Books Written by Charles Darwin

Year	Book Title
1839	Journal of Researches into the Natural History and Geology of the Countries Visited during the Voyage of the H.M.S. Beagle
1842	The Structure and Distribution of Coral Reefs
1859	The Origin of the Species by Means of Natural Selection
1862	Fertilization of Orchids
1868	The Variation of Animals and Plants Under Domestication
1871	The Descent of Man and Selection in Relation to Sex
1872	The Expressions of the Emotions in Man and Animals
1875	Insectivorous Plants
1877	The Different Forms of Flowers on Plants of the Same Species
1879	The Movement of Climbing Plants
1881	The Formation of Vegetable Mold, through the Action of Worms, with Observation on their Habits

FIGURE 15.1   Charles Darwin in 1874 (Courtesy of Elliott and Fry via Wikimedia Commons).

enjoyable to read, but I occasionally employed Google Earth on my iPad to trace his locations and I used my dictionary app to find the meaning of some of his technical words.

If you want to read any of his books, several websites offer the books online free to the public. Be prepared to adapt to his style. We live about 150 years later than Darwin and our modern language is geared to work in a faster-paced society. Today, we strive to get to the point quickly and avoid wordiness lest we lose our audience.

15.3   PRELUDE TO DARWIN'S THEORY OF EVOLUTION

15.3.1   DARWIN LIVED IN A RELIGION-DOMINATED TIME

Of course, the name "Charles Darwin" is most commonly associated with his famous "theory of evolution." This theory was revolutionary when it went public. It contrasted directly with the beliefs of his day. Darwin lived at a time in history when religion rather than science answered the big questions like "Where did we come from?" or "How did the plants and

animals come to be the way they are?" Religionists found the answer in
Genesis of the Bible. God created the plants and the animals, and they
were thought to be immutable. That is, they were designed to be perfect and would never change. Darwins' theory challenged the concept of
immutability and began a debate that is still active to this day.

15.3.2   REJECTING IMMUTABILITY

The problem with immutability is that the fossil record tells an entirely
different tale. Geology was one of the new sciences under development
at this time. It was becoming evident that the Earth's surface consists of
a series of different strata. The older strata lie beneath the newer strata.
Fossils within these strata are uniquely different from each other, and their
very existence is evidence that different kinds of animals lived at different
times in the past.

As further input to his mind, Darwin had observed and reported on
natural science discoveries during his five-year sea voyage around South
America. It was in the Galapago Islands that his theory of evolution began
to gel. Each island had its own unique plant and animal life. That seemed
odd to young Darwin and got him pondering its cause. Darwin was not
alone in seeing this phenomenon in island groupings.

15.4   VOYAGE OF THE BEAGLE

Let us take a deeper look at the experience that opened young Darwin's
eyes to the secret of evolutionary change. Charles had just completed his
BA degree at Cambridge University, when the opportunity of a lifetime
was offered to him. He was offered a "round the world" voyage on a British survey ship, the HMS Beagle. The offer came from Captain Robert
Fitz-Roy, who wanted a naturalist on board to collect, observe, and make
note of anything of interest to natural history.

Charles Darwin's father was initially against the idea, but was brought
around by his brother-in-law, Josiah Wedgewood. This change of mind was
fortunate for posterity because if Charles Darwin had pursued the ministry
instead, he would never have made the same impact on the thinking of

the world. The Beagle's voyage lasted from December 1831 to October 1836, and most of that time was spent navigating in South American seas. Darwin the naturalist spent more time on land than at sea. Captain Fitzroy would drop him off and pick him back up days or weeks later.

Young Darwin had great adventures, which he told about in letters to his sisters. He rode with gauchos, observed southern hemisphere stars from the heights of the Andes, admired the senoritas, and pondered the formation of coral islands as he swam in their lagoons. Yet what he mainly did was collect thousands of specimens and forward them back to England in barrels and crates. His specimens included plants, animals, rocks, and fossils.

15.5   THE GALAPAGOS ISLANDS

The Galapagos Islands (see Figure 15.2) are credited with being the place where Darwin witnessed speciation in action in 1835. He saw the cormorant with shrunken wings evolving into a flightless bird. He saw the black marine iguanas sunning themselves after diving to the submerged coastal rocks to gnaw at the algae growing there. These animals are only found here in the Galapagos and are believed to be descendants of land iguanas from the continent. Dislodged many millennia ago by severe storms and transported on floating vegetation, the pioneer iguanas had nothing to eat on these newly formed volcanic islands except coastal algae. Over the generations, they evolved into sea reptiles capable of staying submerged for 30 minutes or more. However, it is the island finches, that get the credit for Darwin's realization that species evolve. The plant life varies from island to island, and the finches have adapted differently to the different islands. Size and shape of beak is especially varied. By the way, Alfred Russell Wallace would have announced his similar theory of evolution ahead of Darwin if circumstances had been a little different. He was also alerted to speciation by observing the island animals. What was different was that Wallace's islands were in the Malay Archipelago.

Incidentally, I was fortunate enough to tour some of the Galapago Islands on a series of guided land tours from a small cruise ship. It was one of the most unforgettable experiences of my life. I saw marine iguanas

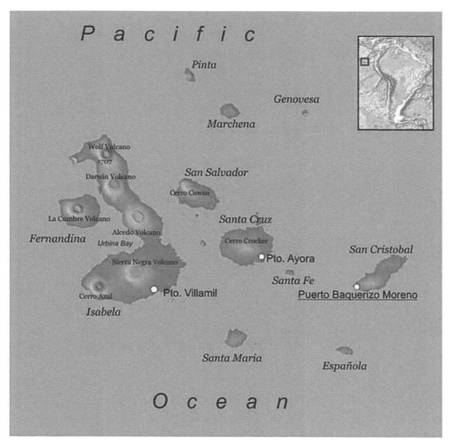

FIGURE 15.2   Map of the Galapago Islands (Courtesy of Wikimedia Commons by Daniel Feher freeworldmaps.net (freeworldmaps.net) [CC BY-SA 2.5 (http://creativecommons. org/licenses/by-sa/2.5)], via Wikimedia Commons).

sunning themselves and sneezing to expel salt that they had swallowed. I saw flightless cormorants with their stubby wings, pink flamingoes wading in an inland lake, small penguins near the shoreline, blue-footed boobies diving in unison into the sea pursuing a school of fish, and male frigate birds with inflated red chests, luring females to their newly constructed nests. I saw wild animals with no fear of man let me approach so close that I could have stepped on them. It was a thrill to follow in Charles Darwin's shoe steps, albeit 160 years later.

15.6 ARTIFICIAL, NATURAL, AND SEXUAL SELECTION

15.6.1 NATURAL SELECTION

Darwin, like other scientists of the time, knew that plants and animals change over time. He had pondered the question of how that was possible and what drives the process for decades before finally announcing his theory to the world. Darwin understood that the species become ever better at survival through a process, which he called "natural selection." The fittest animals survive and pass those superior genes forward into the next generation. It is important that two things be true for natural selection to work. First, the species member must survive through his or her adult reproductive phase, and second, the species member must mate and reproduce. Those superior genes need to be passed from generation to generation. It is always true that species members are different from each other in many different traits. What is important is that those with the best traits for survival pass their superior genes forward into the future generations. The weakest are culled when those with inferior traits fail to either survive or to reproduce. Their genes are lost to future generations. In his writings, Darwin talked about not just natural selection but about two other kinds of selection that shape species. They are artificial selection and sexual selection.

15.6.2 ARTIFICIAL SELECTION

Artificial selection is the term Darwin assigned to man's tinkering with species. Darwin, himself, was fascinated with botany and breeding exotic forms of pigeons. Long before humans had an inkling of how reproduction and inheritance works, they were changing species of plants and animals. Agriculture began in the Fertile Crescent of Iraq around 10,000 years ago. There was a lot of experimentation that went on in the quest to domesticate and optimize crops. Selecting the largest seeds for next year's crop is a form of artificial selection. Over many seasons of planting only the largest seeds, the normal crop becomes one of the larger seeds. Corn was developed in America over the course of 5000 years. Although originally a grass with small seeds, it became a vegetable with large kernels. Crossing species of plants is another tool of artificial selection. The resultant hybrid plants, so obtained, had have advantages to the farmer over the parent species. Flowers have also been a rich area for hybrid tinkerers. Here the goal is artistic appeal; the art of producing rich combinations of colors and patterns.

Animals have also been the subjects of artificial selection adjusting. Dogs were developed from wolves thousands of years ago by means of artificial selection, and they have become a valuable partner to the hunter-gatherers in a variety of endeavors (e.g., tracking prey, retrieving game, guarding against intruders, etc.). Wolf pups today grow up to be overly aggressive adults and are unsuitable as pets. So there must have been a selection process that retained and bred the tamest adults and disposed of the wilder wolves. Today, we have over 500 different breeds of dogs in the world, and they are all of one species. Artificial selection led to optimized domestic animals of many types. Dairy cows are very different from beef cattle, as an example.

15.6.3 SEXUAL SELECTION

Sexual selection may have fascinated Darwin more than natural selection. He wrote more words about it in his book, *The Descent of Man and Selection in Relation to Sex* than he ever wrote about natural selection. Remember that natural selection is nature's device for maximizing a species adaptation to the environment in which they live. The fittest get to pass their superior genes into future generations. However, the fittest individual can make no contribution to optimization of the species if he or she never mates and have offspring. That is where sexual selection comes in. What governs the process of being able to mate? What about finding the best mate possible? It is advantageous to find the best mate possible to give the offspring a leg up in the competitions for survival and mating that they will face.

Grazing animals usually adopt a harem-type arrangement with one dominant male servicing all the females in his harem. This leads to violent competition to determine which male will be that dominant male. Yes, sexual selection can be a deadly event. Deer and elk grow antlers for the competition. Mountain goats have curved horns for ramming encounters.

There are yet untold tales of how these weapons of combat evolved. Birds have a different kind of social system for mating. They usually form male-female pairs by a process in which the female does the selecting from a group of courting males. Female birds are usually dull-colored and camouflaged to protect them from predators. Males are the brightly colored, fancy-feathered gender of the bird world. The male bird evolved this brazen display against his interests of avoiding predators. In other words, sexual selection can be a more powerful evolutionary force than natural selection. Birds that live in forests have difficulty seeing very far due to the foliage and have developed a solution to the problem; they sing. So, part of sexual selection evolution is being able to sing the correct song.

How about man? Darwin believed that sexual selection was a major factor in shaping man. Unlike the chimps and bonobos, we do not have indiscriminate sex and leave the burden of raising the child to the female alone. We form pair bonds; often lifelong pair bonds. Perhaps this is why the human female is the more ornate gender of our species.

15.7   SHOCKING THE WORLD

15.7.1   ORIGIN OF THE SPECIES

Darwin only first considered the transmutation of the species in March 1837, five months after arriving back in England. The five-year journey aboard the HMS Beagle had provided experiences that got him thinking deeply about it. While in Uruguay and Argentina, he encountered fossils of extinct animals (e.g., armadillo and rhea). He also noticed that somewhat similar animals were alive there. He pondered whether there was an ancestral relationship between them. However, the experience of visiting many of the Galapago islands and collecting plant and animal specimens was even more important. He really thought it curious that each island had its own unique set of flora and fauna. He particularly noted that finches of the islands had distinct traits. Their beaks varied in size and shape in adaptation to the available plants of the different islands.

Darwin formulated his theory of evolution over the two years following his return from the voyage, but did not publish it. Perhaps he realized how controversial the book would be and decided to build a rock-solid case before going public. Instead he kept his research under wraps and confided in only a small group of trusted friends. He operated this way for over 20 years; then something happened that forced him to make a decision about publishing. Alfred Russel Wallace had arrived at the same conclusions as Darwin and had sought Darwin's counsel. Darwin's friends urged him to publish his theory lest credit he deserved be lost. He did publish while giving Wallace his due credit.

In 1859, Charles Darwin published his book, *The Origin of the Species by Means of Natural Selection*. It was an instant bestseller and was as controversial as Darwin feared.

15.7.2   A PROBLEM WITH THE THEORY

Charles Darwin had one very serious problem with his theory of evolution; he didn't know how the mechanism of inheritance actually worked. Of course, everyone knew that children looked very similar to their parents, but no one knew how it worked. If traits of the mother and the father were somehow averaged so that the offspring had intermediate traits between those of the parents, the full survival value of a superior trait was compromised. That would be a problem for Darwin's theory. In fact, he was challenged by Scottish Professor Fleeming Jenkins on that very weakness and he had no response. Strangely, the answer to the puzzle existed at the time, but Darwin never knew of it. The mechanism of inheritance was being discovered by a German monk named Gregor Mendel concurrent with the publication of Darwin's theory of evolution. An indivisible unit of inheritance (i.e., the gene) was passed from parent to offspring in an unadulterated form. In a twist of fate, Darwin never learned of it.

15.8   DARWIN'S LEGACY

Darwin's theory of evolution is the foundation of the biological sciences and has led mankind to an age where disease and suffering have been vastly reduced due to advancing technology. Our understanding of the living world began with Darwin's legacy. In the next chapter, a series of scientists moved it further along. It is called the Modern Synthesis.

15.9 SUMMARY

Charles Darwin was a man of the nineteenth century. He was born in 1809 and died in 1882. He lived in a time when few questioned the biblical account of human origins, and yet he was destined to shake that belief system up like never before. Darwin was in search of his life's work in the early years, but the chance to participate in a naval exploratory journey was fortunate for him because it determined that he would spend his life in the arena of biological science. In Darwin's time it was believed that species of plants and animals were created by God in a perfect form and had never or would never change. Living things were immutable. However, some scientists were questioning this belief, and Darwin saw the situation more clearly than others, especially after his voyage on the HMS Beagle. In particular, Darwin's observations of plants and animals on the Galapago Islands off the coast of Ecuador suggested that these living things adapt to the environments they find themselves in and are altered by this process.

For many years, Darwin avoided publishing his theory of evolution by natural selection. He had undertaken numerous studies of changes in plants and animals by artificial selection, whereby man influences the resultant traits by selective breeding. Natural selection was the same process except nature was determining the outcome, not man. Darwin may never have published it at all, except for the fact that Alfred Russell Wallace had undergone the same kind of island epiphany in the Malay Archipelago. He had also concluded that natural selection accounted for the differences between plants and animals from one island to another. Darwin's friends persuaded him to publish and claim credit for those years of work on this thesis, so in 1859, Darwin published his famous book, *The Origin of the Species by Means of Natural Selection.*

As important as that revolutionary work was, it was far from the end of Darwin's scientific contributions to the world. In 1871, he published *The Descent of Man and Selection in Relation to Sex*. The power of sexual selection in modifying animals was not warmly embraced until recently by the scientific community, and yet it is a major driver of evolutionary change. Darwin investigated many related topics and wrote books. His rich contribution is still not understood by most people in the world.

CHAPTER 16

THE MODERN SYNTHESIS

CONTENTS

16.1 SCOPE

Darwin's theory of evolution came from his great insight into how plants and animals change as they adapt to changing environments and new threats to their survival. His analysis became the new foundation of biological science in the nineteenth century. However, Darwin never resolved the vexing question of how superior traits of an individual pass to his or her offspring undiluted. At the time, it was believed that traits of the offspring were a blend of the traits of mother and father. If that supposition was true, a blended trait diminished natural selection. However, if traits could pass to offspring unaltered, natural selection was a powerful force.

The ever-advancing accumulation of scientific research in the latter nineteenth and early twentieth century eventually answered this question and much more. The concept of an indivisible unit of inheritance called a gene was demonstrated and characterized. The science of the gene, i.e., genetics, is central to our understanding of the modern evolutionary theory. You will see aspects of genetics frequently in the following chapters. Although understanding human origins is our primary theme in this book, genetics is a complementary theme of great importance, and it is an important tool in reconstructing the past.

16.2   MENDEL WAS A CONTEMPORARY OF DARWIN

Darwin published his book *The Origin of the Species by Means of Natural Selection* in 1859 without knowing the mechanism of inheritance. If he had been aware of Gregor Mendel's research going on in Germany, he could have amended his book and eliminated the greatest weakness in his theory. Unfortunately, Darwin never learned of it, and many decades went by before the genetic inheritance became a part of the theory of evolution in what has been called the "modern synthesis."

Gregor Mendel (1822–1884) was a German monk whose extensive research with hybridization of peas yielded the first clear understanding of how inheritance works. Between 1856 and 1863, he planted over 29,000 pea plants having distinctive traits. For example, some of Mendel's peas had either yellow or green color and either smooth or wrinkled skin. He followed the plants over many generations and observed their inheritance traits. He discovered that the offspring from mating different parent plants resulted in one pure recessive, two hybrids, and one pure dominant. Mendel postulated that there was a unit of inheritance called a "gene." The gene was responsible for the trait being observed.

It is too bad that Charles Darwin never learned of Mendel's important discoveries. The concept of the gene solved the problem of how a superior trait might be inherited intact. Otherwise, it was assumed that sexual reproduction averaged all of the traits of the father and the mother. It was difficult for Darwin to see a resolution to the problem if averaging occurred. An individual might have superior traits for survival, but what good did that do for the offspring if those superior traits were diluted by averaging? Mendel's genes allowed the superior trait to be inherited intact.

Mendel's heredity research on plants also yielded other important conclusions. He determined that genes come in pairs, one from the male and one from the female parent. Although those parents each had pairs of genes, each parent only contributed one of each gene pair to a given offspring. In other words, there is a 50% chance that an offspring inherits a particular gene from their parent. So, if the parents' two matching genes produced different traits, like blonde hair or red hair, there was a 50% chance that the offspring would inherit red hair, for example. If many offspring who had inherited a gene for blue eyes from one parent and a gene for brown eyes from the other parent, all had brown eyes, then we say the gene for brown eyes is a dominant gene and the gene for blue eyes is a recessive gene. Mendel discovered that genes can be either dominant or recessive from his plant research. He also learned that for recessive genes to be expressed, both of an individual's genes must be recessive.

Mendel published his paper on pea inheritance traits in 1866. He continued his research with fuchsias, maize, and other plants. Mendel was able to mathematically predict from his research how traits are inherited. He did the original pioneering work developing the laws of genetics.

Now, let me tell a personal story about inheritance. I believed that all four of my children inherited my blue eyes rather than their red-headed mother's green eyes. However, one day my youngest son told me that he had different colored eyes. I hadn't noticed it before, but when I carefully compared his two eyes, there was indeed a subtle color difference between them. His eyes were actually a blue-green color with one eye slightly bluer than the other. So, all of us can examine the effects of inherited traits in our daily lives. Hair color distribution was interesting too. My oldest son was blonde like me, but my other three children all had strawberry-blonde hair, with some redder than others. Their mother had fair skin that easily burned from sun exposure rather than tanning, whereas I could get a summer tan if I was careful about exposure. So, from observing my own children, I noticed a correlation between red hair and pale skin. The redder the hair colors of my children, the poorer their ability to tan. By the way, genetics is usually not as simple as in Mendel's experiments. He was lucky that one gene controlled one trait. The usual case is that several genes control an observable trait. The human eye color is governed by 15 different genes.

16.3   GENETICS LOST AND REDISCOVERED

Despite the long years devoted to genetics research and the great discoveries that he had made, Mendel's discoveries never caught on in the scientific community during his lifetime. His discoveries were not to die though; they were rediscovered in the early twentieth century. I learned from Matt Ridley's book *Genome* that Mendelian genetics was independently rediscovered in 1900 by three different botanists. Each of them (Hugo de Vries, Carl Correns, and Erich von Tschermak) duplicated Mendel's work on different species, unaware of Mendel's earlier contribution. From that point of rediscovery, the scientific world was fascinated with Mendelian genetics. The particulate nature of the theory was compelling. Thomas Hunt Morgan was one its strong proponents and founded a school of genetics, which included the chromosomal role in inheritance.

16.4   EVOLUTIONARY GENETICS AND FRUIT FLIES

What really kick-started evolutionary genetics was the discovery that genes can be altered through mutations, along with an experimental method of studying the phenomena. Mendelian genetics explained inheritance but didn't explain how evolutionary change might occur. Mutations were the key. Hermann Joe Muller won a Nobel Prize for showing that genes are artificially mutable. Mutation experiments have told us an immense amount of information about biological science. Fruit flies have been the favorite insect in mutation experiments because they reproduce so rapidly, and many generations can be examined in a short period of time. Muller discovered that he could induce mutations in fruit flies by subjecting them to x-rays.

16.5   CHROMOSOMES

Some breakthrough discoveries on inheritance began to be made in the 1880s. Chromosomes were discovered through microscopic examination of cells and are found in all living things. They play a vital role in inheritance. Although some primitive life forms do not have a nucleus in their cells, most animals do have a nucleus, and that is where an individual's chromosomes are found. Chromosomes exist in pairs, which is the end result of sexual union between male and female. In the case of humans, we have 23 pairs of chromosomes or 46 chromosomes total. One of those chromosome pairs determines our gender. There are X type and Y type sex chromosomes. If the pair consists of two X chromosomes, you are a female. If you have an X and a Y chromosome, you are a male.

Different species may have more or fewer chromosomes than humans. Apes (e.g., gorillas and chimpanzees) have 48 total chromosomes, which is close to our 46. The plant rye has 14, and the plant maize has 20. Your pet cat has 38, but your pet dog has 78. The kingfisher bird has 132, and a garden snail has 54. A fruit fly has a measly 8 chromosomes, but the silkworm has 56.

Chromosomes are made of deoxyribonucleic acid (commonly known as DNA), a wondrous molecule capable of carrying information of biological importance. Everyone today knows that DNA is unique to an individual and is used commonly in police work to identify a perpetrator or clear an accused person of a crime. It is also used to prove parenthood or to learn facts about your ancestry. DNA is the code-containing material in our genes, and because that coded sequence is different from gene to gene, individual genes can perform divergent functions. So, what is the relationship between genes and chromosomes then? Each of our chromosomes contains many different genes. For example, Chromosome 1 has 2000 genes within it, whereas chromosome 18 only has 200 genes. Our sex chromosomes are quite different from each other in this sense: The Y chromosome only has 50 genes, whereas the X chromosome has 800 genes.

16.6   DNA COMPLETES THE PICTURE

In Part IV of this book, we will take a more detailed look at what science has learned about DNA and its role in inheritance. We will learn that DNA is a double-stranded giant molecule that contains, among other things, the 21,000 different genes that we humans inherit from our parents. We will learn that our genes contain a four-letter code that instructs our cells how to build a particular protein from the 20 amino acids common to living things. These proteins are essential in reproducing our own kind and in maintaining our bodies day-to-day.

The science of DNA and living things is still in its infancy, and new discoveries are being made daily. Yet by the 1930s, enough had been learned for scientific pioneers to marry Darwin's theory of evolution with modern genetics.

16.7   THE MERGER OF EVOLUTION AND GENETICS

A lot of new scientific facts about inheritance have been learned since Darwin's time. Darwin's insight and accomplishments are all the more impressive when you realize that he knew nothing about the mechanisms of inheritance. We now have a pretty clear picture of the way unaltered traits from the mother or from the father are inherited intact by the embryo. At the time of conception, genes are acquired from both the father's sperm and the mother's ovum. The embryo thus acquires some traits from the mother and some from the father. Gregor Mendel had worked out some of the inheritance rules from his study of peas during Darwin's time, but that discovery never made it to Darwin or his cohorts. In the early twentieth century, genetics was rediscovered as well as chromosomes, the structure of DNA, and the biochemistry of organisms. It became time to add all that modern knowledge to Darwin's theory and have an improved, up-to-date theory. This synthesis is called the "Modern Synthesis."

16.8   SUMMARY

Charles Darwin had a problem with his theory of evolution in that he didn't know how traits were passed from parents to their offspring. If the traits of mother and father were blended or averaged in the offspring, that put his concept of natural selection at risk. While he was writing *On the Origin of Species* in England, a German monk was running experiments with hybrid plants that could have solved the problem for him. That monk, Gregor Mendel, determined that there was an elementary unit of heritance that carried the trait of a parent intact to the offspring. This unit, called a gene, saved natural selection but Darwin never would learn of it.

Over the following decades, the modern picture of inheritance began to unfold. Microscopic examination of the nucleus of cells revealed new information. Structures called chromosomes were seen and they appeared in pairs. The process of cell division was observed, and the chromosomes divided themselves into two groups. New partners for the chromosomes were built. Chemical analysis of the chromosomal material showed that it was nucleic acid. Later, it would be shown to be deoxyribonucleic acid, or DNA for short. The genes proposed by Mendel now had molecular significance. They were located within the chromosomes themselves.

Scientists also learned more about how genes affect biological processes like embryology. Fruits flies became a favorite test animal for genetics researchers because of their very short reproductive cycle. Many generations of fruit flies could be observed over the course of a few weeks. Much was learned about how bodies are assembled from a small embryo and how traits are inherited from these studies.

Now we know that DNA is a very special molecule. It contains the genetic code that instructs the building of an embryo into a baby and programs its childhood, adolescence, and other stages of its life. It also instructs the body on how to build replacement cells, hormones, and other biochemical entities. And so Darwin's original theory was merged with the newer scientific discoveries about inheritance. The resulting modified theory is called the modern synthesis.

CHAPTER 17

RICHARD DAWKINS, THE DARWIN OF OUR TIMES

CONTENTS

17.1 SCOPE

Richard Dawkins is one of many evolutionary theorists alive today. Yet, I think he has been the most effective of them in understanding the complexity of this part of biological science and presenting its concepts to the world in both an interesting and a digestible manner. He, like Darwin, has contributed an impressive list of readable books that teach evolutionary concepts. His writing style makes his topics interesting because he has a wealth of animal behavior tales at his disposal to weave into his narrative. His writing style is enjoyable because he is imaginative in breaking down the complexity into chunks of everyday concepts, which we can follow, and then translating each lesson's essence back into the sphere of biology.

Dawkin's entry into the public limelight came with the publication of his book, *The Selfish Gene*, where he argues that the main character in the evolutionary story of life on Earth is the gene, not the individual. We individuals are but temporary vessels for our nearly immortal genes. Dawkins stretches our concept of the gene and its powerful influence. Our genes not only determine how our bodies assembled as embryos, grew as infants, adolescents, and adults, but also determine our behavior. In fact, our genes can even determine the behavior of others. I was skeptical as I first read these revelations about gene power, but Dawkins argues for his theories skillfully.

17.2   RICHARD DAWKINS

I tend to see Richard Dawkins in many ways as the Charles Darwin of our day and age. Dawkins advocates many of Darwin's evolutionary principles such as incremental change and gradual change over many generations. However, Dawkins has the advantage over Darwin in understanding how inheritance works. Dawkins has mentally absorbed the post-Darwin evolutionary discoveries, collectively known as the modern synthesis. Thus, Dawkins has been able to incorporate Mendelian genetics, discovery of chromosomes, discovery of DNA, and the biochemistry of protein production via genetic coding into his knowledge of evolutionary processes. Consequently, Dawkins is the spokesman for Neo-Darwinism in our times.

Darwin and Dawkins both made their mark as insightful evolutionists, so they are similar in intellect, career purpose, and commitment. But there are notable differences in persona between these great men. Dawkins is an unabashed and vocal atheist, whereas Darwin tended to suppress his religious skepticism. Moreover, Darwin was very reluctant about making statements that might offend his religious wife. There are also differences in assertiveness. Dawkins is openly combative in defending his positions on evolution, whereas Darwin avoided conflict and let others do his fighting for him.

Those differences aside, the two men have a lot in common. They are both Englishmen although Dawkins was born in Kenya and lived there until he was eight years old (Figure 17.1). And without a doubt, they both rank high as prolific scientific authors and educators. Finally, they present a continuous evolutionary philosophy. Darwin originally arrived at the concept of natural selection and evolutionary change. Dawkins absorbed the concepts of the modern synthesis and became the primary spokesman for Neo-Darwinism. One thing that makes both of them especially interesting is their extensive knowledge of animal behavior and their skill in finding the threads that tie the whole complex story of life on Earth together with simple yet universal explanations.

Dawkins is an ethnologist, which means he is an expert on animal behavior. His wide knowledge of animals has enabled him to make his books delightful reads by weaving in numerous animal stories and unusual facts. However, the idea that has made Dawkins most famous is his concept of the selfish gene. Most other theorists had selected the species or individual member of a species as their focus in evolutionary processes. Richard Dawkins argued that this focus is incorrect and has instead established the gene as the principle unit of inheritance. Unlike the individual member of a species, whose time on Earth is short, the gene may persist for thousands, even many millions, of years. Moreover, the gene is stable in its identity across many generations, whereas the individual is a unique

FIGURE 17.1   Photograph of Richard Dawkins (Courtesy of Wikimedia Commons by Cstreet (Christopher G. Street), Bransgore, Dorset, England, UK. (Own work.) [Public domain], via Wikimedia Commons).

combination of traits, which will not be duplicated in his offspring due to the genetic mixing from sexual reproduction.

Table 17.1 shows the list of Richard Dawkins' books that relate to evolution. A person can learn a lot about evolutionary theory from them. They are written for the scientifically curious reader with explanations suited to the novice. Moreover, his books are essentially free of the heavy technical jargon that is commonly encountered when evolutionary biologists communicate directly with each other.

17.3 THE ULTIMATE COMMON ANCESTOR

Charles Darwin believed that all living things descended from a common ancestor and that each creature alive today has an unbroken chain of descent tracing back to that first common ancestor. We are talking about hundreds of millions of years and, in our case, ancestors that were apes, prosimians, proto-mammals, reptiles, amphibians, fish, and proto-vertebrates. I thought that was an incredibly bold conclusion considering the primitive state of biological knowledge in Darwin's day. However, Richard Dawkins took the concept a lot further. In fact, he wrote an entire book to show how various life forms are related. The book, *The Ancestor's Tale*, is written in the style of Chaucer's *Canterbury Tales* where each of the travelers gets to tell their individual tale. Instead of the Miller's tale or the Nun's tale, Dawkins

TABLE 17.1 Books on Evolution by Richard Dawkins

Year	Book Title
1976	*The Selfish Gene*
1982	*The Extended Phenotype: The Long Reach of the Gene*
1986	*The Blind Watchmaker: Why the Evidence of Evolution Reveals a Universe Without Design*
1995	*River Out of Eden*
1996	*Climbing Mount Improbable*
1998	*Unweaving the Rainbow: Science, Delusion and the Appetite for Wonder*
2004	*The Ancestor's Tale*
2009	*The Greatest Show on Earth*
2013	*An Appetite for Wonder*

has substituted the Ape-Man's tale, the Gorilla's tale, the Dodo's tale and many, many others. The book begins with modern animals but moves progressively farther and farther back in time until we are in a time where only one-celled animals lived. The 614-page journey must have been an ambitious endeavor, requiring a laborious research effort to complete.

17.4   THE SELFISH GENE

Living things begin life, have a childhood, reach adulthood, reproduce, get old, and die. The life of an individual is fleeting, whereas the entity that seems to be immortal is the gene. Richard Dawkins became famous overnight with his first book, The Selfish Gene. His thesis was that the proper unit of evolution is definitely not the species, not even the individual, but most definitely should be the gene. It is the gene that struggles to perpetuate itself into future generations. The proof of a gene's success is its frequency in the world. In order to become dominate, a gene must outcompete rival genes. It must act selfishly.

What is a gene? For now, think of a gene as a discrete unit of inheritance responsible for a specific trait. The trait might be blue eyes, keen hearing, muscular strength, and others. Later in the book, we will see the identity and behavior of a gene from a biochemical perspective. However, we don't need to know about that yet. Now while it is true that a single gene sometimes controls a given trait, it is more likely that several genes work in concert to control a trait. Humans and other animals have thousands of genes in their genome. Twenty thousand was the last number I heard for humans. It was previously estimated to be much higher. Genes actually come in pairs. We get two copies of each gene, one from our mother and one from our father. Despite the fact that a gene is often working in concert with other genes, one particular gene may make the critical difference that tips a trait into the direction that makes an important difference to the survival or reproductive prowess.

17.5   THE INFLUENCE OF JOHN MAYNARD SMITH

Richard Dawkins was drawn to the mathematical side of biology as a college student and developed a fascination with the power of computers when applied to biological and evolutionary problems. Indeed, he became an accomplished computer programmer and created many of his own programs and tools. I admired this interest of his, which he made a feature in several of his books. For example, he illustrated computer-generated life forms. Although I never reached Dawkins' level of computer skill, I wrote several of my own programs using the Pascal programming language. I created computer denizens, which could mate, have offspring, be predator or prey, and other traits. It was fascinating to watch generations whiz by while populations diminished or rose so fast that they crashed the computer.

It turns out that evolutionary principles can be tested using mathematical modeling with rewarding results. Dawkins was drawn to this practice under his mentor John Maynard Smith, who wedded game theory modeling to biological studies. Dawkins applied these concepts to his world of the selfish gene. In his book, *The Selfish Gene*, Dawkins applied game theory to a population of doves and hawks. That is, an imaginary world of individuals is visualized where the members are either doves or hawks. Hawks are aggressive individuals who excel at conflict, whereas doves fight poorly and avoid conflict. Dawkins applies a numerical rating system where he awards 50 points for a win, 0 for a loss, −100 for being seriously injured, and −10 for wasting time in a long contest. Then various starting point scenarios are evaluated such as: all doves, all hawks, 50% hawks and 50% doves, etc. There is great short-term benefit to be a hawk in an all dove environment, but that changes as the percentage of hawks increase because the penalty for being seriously injured is severe and the odds of that happening increase in a hawkish population. It turns out that there is an ideal population mix of hawks and doves. Dawkins calls it the "evolutionary stable strategy" or ESS.

These basic concepts can be applied to a vast number of biological situations. Dawkins ties the results to the success of individual genes. The individuals exist only briefly as the generations roll by. It is the genes that are the permanent players in these simulation studies.

17.6   THE INFLUENCE OF W.D. HAMILTON

Dawkins tells us in *An Appetite for Wonder*, that his ideas for the *Selfish Gene* book were heavily influenced by concepts developed by W. D.

Hamilton. These ideas, collectedly called "kin selection," are an integral part of the modern synthesis discussed in Chapter 16. The gene becomes the central entity of evolution. Its success in passing itself forward into future generations and even increasing its presence can be measured and predicted. Other genes, being less successful, will decrease or even disappear over time. Sometimes animals behave in an altruistic manner. For example, bees give their lives defending the hive against invaders. Hamilton's Rule predicts when such behavior will or will not occur based on the cost/benefit ratio and relatedness of the participants. Hamilton was personally very interested in how altruism factors into the behavior of social insects such as bees, ants, wasps, and termites. Many of the ants in a colony are sterile and have no chance of having offspring themselves. Yet they share genes with their fertile brothers and sisters and do have motivation to see those genes passed forward.

17.7   THE GENE AND EVOLUTION

Later in this book, we will see the gene viewed from different perspectives such as: it being a string of nucleotides, being a coded instruction to assemble specific proteins, being an entity with a very long ancestral history, or as being a part of our DNA, sometimes capable of duplicating and jumping around. However, Richard Dawkins focuses our minds on something different about genes. He focuses upon the gene's function as a unit of inheritance and as an entity with its own will and influential power at a distance.

Through Dawkin's eyes, genes attempt to influence their surroundings with the goal of increasing their population in the host bodies of future generations and assuring their survival however they can. He provides you with countless examples of this actually happening in the world around us. This perspective on genes is the essence of his numerous books.

In my opinion, genes are responsible for the most awesome things we observe in our lives. Human babies crawl on the floor, but one day, they gleefully stand up and attempt to walk. They babble as tots, but soon can talk to us in two-word sentences. Before long, they can speak clearly in full sentences. How can this be? No other animal can communicate like this, yet babies are preprogrammed to learn their native language at an early age. In a different example, consider wildebeests on the Serengeti plains. Their newborns struggle to rise to their feet and stand on shaky legs soon after birth. They must learn to walk, then to run rapidly as soon as possible if they are to survive. Predators will dine on the bodies of the slower learning ones. It is an amazing thing and illustrative of evolution as a determinant in animal behavior.

17.8   GENES AND GAME THEORY

Dawkins uses the language of the game theorists, where genes appear to be acting in their own interest and various strategies either succeed or fail. However, genes do not think or even have a brain. They have no actual strategy, which is simply a concept of convenience for the analyst. What genes actually do is duplicate themselves during the reproductive process and get passed ever forward through time if they contribute to the survival and/or mating of the individual bodies they occupy. The concept of a single gene determining its own destiny is also a conceptual device. Most of the time, traits are determined by several genes. In this case, our singular gene may influence the success or failure of the team of genes.

Genetic predisposition is not the only factor in real-world events; luck plays a role too. For example, that newborn wildebeest may get eaten by a lion regardless of how fast he develops into a runner. The wildebeests are programmed genetically to give birth within a short time. The predators are feasting on the young, but they can't possibly eat all the newborn wildebeests; there are too many. Bad luck aside, good genes make a difference to those baby wildebeests spared when most vulnerable. Genes for a quickly acquiring fast running ability get passed to future generations, whereas those for slightly slower acquisition of this trait do not. Human genes for upright walking on two legs or genes for talking and comprehending speech have developed as unique traits for humans as is clearly evident as we observe our babies develop. Genes are important to understanding our origins.

17.9   THE EXTENDED PHENOTYPE

"The Extended Phenotype" is the name of Dawkins' second book, and it is essentially the sequel to *The Selfish Gene*. So what is a phenotype? If you look the word up on the Internet, you will get a precise and very technical biological definition. I tend to think of it as an organism's characteristic behavior. The words genotype and phenotype are often used together, where genotype refers to the organism's heredity and the phenotype is the intended effect. A person's blue eye color is an example of a phenotype.

Dawkins' concept of an extended phenotype has to do with genes influencing behavior outside of the body of the gene-carrier. The characteristic nest that robins build is an extended phenotype of robin genes. Beaver dams, spider webs, and gopher tunnels are extended phenotypes too. The phenotypic effects of a gene had been defined as all of the effects that a gene has on the body in which it resides. Dawkins changed that definition to "the phenotypic effects are all the effects that a gene has on the world."

17.10   SUMMARY

If the question is "Who is the modern-day Darwin?" then the answer has to be Richard Dawkins. Like Darwin, he has been a major contributor to evolutionary biology. However, unlike Darwin, he has a special talent for communicating evolutionary science to the general public in a compelling manner. Dawkins' work has focused on the gene as the major player in the story of evolving species. He is widely known for his book, *The Selfish Gene*, which encapsulates the theories and findings of other scientists such as W.D. Hamilton, John Maynard Smith, and others. The main idea is that the genes are long-lived entities whereas the plants, animals, and even human bodies they occupy are merely temporary vessels that do their bidding. Successful genes make many copies of themselves as they move forward in time generation by generation. Some are nearly eternal entities. Unsuccessful genes disappear. Genes can even project their will at a distance as in the building of beaver dams or spider webs.

CHAPTER 18

MECHANISMS OF SPECIATION

CONTENTS

18.1   SCOPE

The pre-Darwin belief was that species don't change. They were created as perfect entities and always remain the same. However, Darwin showed that species do change. New species also come into being, whereas older species may go extinct. In this chapter, we shall discuss how species arise, change and disappear, and what forces influence those occurrences. These concepts are important to all living things, of course. However, we are particularly interested in how they help us understand the transformations that changed a tree-dwelling ape into a ground-dwelling, upright walking, large-brained human. In order to properly understand these changes, we need to understand the basic principles of evolution. For example, changes do not occur in a species unless those changes benefit the existing species members. There is no such thing as long-term planning. Another principle is that changes can only operate on the existing plant or animal. Bat wings are made from stretched skin and not feathers for this reason.

18.2   WHAT IS A SPECIES?

How is it possible for one species to evolve into another species? Well, let's consider our own species for a moment. There are over seven billion of us on the planet now, and many of us look quite different from each other. North Europeans tend to have fair skin, blonde, red or brown hair; blue, hazel, green or brown eyes; and perhaps freckles. Many Africans have black or brown skin, black kinky hair, and brown eyes. Asians are distinguished by a yellowish-white skin, black hair, narrow-opening eyes, and smooth features. We could go on and on describing the different people of the world by their stereotypical features. Yet, we are all one species. The proof of that statement is that any fertile man and woman in the world can have children together. *That is the acid test of being in the same species.* Thirty to forty thousand years ago, Neanderthals lived in Eurasia with our relatives, and they are considered to be a separate species by some. As we shall see in Chapter 28, Neanderthals and our relatives had children together, and we carry a trace of Neanderthal DNA in our genomes. So perhaps they are sub-species and not a separate species after all.

Our best friend, the dog, has breeds exhibiting greater diversity than we humans do. Yet, they are one species too. Where we can see the likelihood of any human male and female pair being able to mate, it is hard to imagine it happening between a Saint Bernard and a Chihuahua. Conception would take place if it were physically possible for them to mate though. The variety of different dogs in the world is the result of Darwin's artificial selection process. Man has selected for the traits he wants by allowing mating, which builds those traits and disallowing mating that diminishes those traits.

18.3   SPECIES CHANGE ALL THE TIME

18.3.1   OBVIOUS CHANGE

Often it is obvious when a species has changed because it looks different than it did at a previous time. Paleontologists tell us that horses were once much smaller and had three toes instead of one. Moreover, horses and zebras and donkeys are now considered to be separate species. These single-toed animals all descended from a common ancestor, and they can interbreed with each other although the offspring are infertile.

18.3.2   SUBTLE CHANGE

Some species have been around for millions of years, but we cannot be sure they didn't change with time. Usually, all we have to examine is a collection of bone fossils and we can only learn so much from them. The preserved bones of a species may look identical over thousands or even over millions of years, yet the animal may have undergone numerous changes that we cannot see in the bones that remain. Europeans developed immunity to a host of diseases during the agricultural development period. Those same diseases killed nearly 90% of the Native Americans, who had no immunity to them. Yet, skeletal matter would probably not indicate these immunities or susceptibilities.

18.4   GENERAL COMMENTS REGARDING SPECIES CHANGE

18.4.1   EVOLUTION IS LIMITED BY ITS RAW MATERIAL

There are definite limits to how an animal can change during an evolutionary phase. Remember that only minute changes are possible for a given generation. Most importantly, changes to the species are limited by the body, with which we begin. Birds and bats converted their forearms into wings. They did not sprout new wings on their backside and retain their forearms. When we look at the skeletons of different kinds of mammals, we see that each of them is a modification of an original body plan. I had an epiphany in the dinosaur section of Chicago's Field Museum many years ago. I was awed by the immense size of an assembled dinosaur skeleton. I remember becoming aware of how similar my own skeleton is to that of a dinosaur. Although the dinosaur was much larger than me, and most any of his bones would be too heavy for me to lift, we were of a similar body plan. His legs were comprised of a single bone above and two bones below the knee joint. So are my arms and legs. His bones include a spinal column, a skull, a ribcage, and so do mine. So many animals have these same skeletal features that we conclude that they are somehow related.

18.4.2 THERE IS NO LONG-RANGE PLAN

Certain fish evolved into amphibians, certain amphibians evolved into reptiles, certain reptiles evolved into dinosaurs, and certain dinosaurs evolved into birds. It is easy to see a long-range plan for one type of animal to transform into another type of animal. However, scientists think that is not the case at all. Whatever incremental changes are observed in the fossil record must have been of advantage to the animal at that particular time. Natural selection only operates in the here and now. On the other hand, newly acquired body parts can be co-opted for new uses. Feathers are thought to have evolved originally for preserving warmth. Once acquired, they were available for new uses; at first better gliding, and eventually actual flying.

18.4.3 SPECIALIZATION

The more generalized the evolving animal is, the more variations are potentially possible. Once an animal specializes, there are limits to the evolutionary paths open to it. An example follows of a generalized animal specializing.

When the bat was a more generalized rodent, perhaps something like a mouse, it improved its feeding opportunities by gliding through the air. Capitalizing on the airborne feeding niche, the bat grew longer fingers, with webbing between them. This wasn't an overnight process, but one of minute incremental changes each of advantage to the then existing animal.

Countless generations went by where the better gliders flourished, and the poorer gliders perished. Gliding advanced to actual flying. Each step in the process increases the specialization in a body suited to airborne feeding. The bat is a mammal that learned to fly, but a reptile underwent a similar transformation during the Age of Dinosaurs. I am referring to the pterosaurs. Like the bats, they also had an adaptive radiation resulting in numerous flying species. We have their fossils to remind us of their past prominence. Birds are another animal that specialized by converting arms into wings. These specialized animals can no longer revert to quadrupeds. Any future changes must build on that specialized body plan.

18.4.4 DIVERSITY

The individuals within a species may vary considerably from each other. In general, the diversity of older species exceeds that of more newly formed species. That greater diversity is an advantage to the species because it allows the species to better survive threats to its survival. A study of finches in the Galapagos Islands (*Ecology and Evolution of Darwin's Finches* by Peter R. Grant) is a case in point. Occasional dry years reduced the quantity of small seeds available for the finches to eat. There was a larger cactus seed available, but only finches with an adequately large beak could free them from their seed case. The result was a large increase in the percentage of individuals with larger beaks and the disappearance of finches with small beaks. However, when the normal rainfall years returned, the distribution of large and small-beaked finches returned to its original state. On the other hand, if climate change were to produce a permanently drier climate for these finches, the species would permanently change in the direction of stronger beaks.

18.4.5 MUTATIONS

Only so much change can come from individual variability. The big and lasting evolutionary changes come from mutation. A mutation is a chemical change in the DNA molecule. The changes can occur from mistakes in transcription of the code or by random events, like the force of a cosmic ray hitting the DNA molecule. Most mutations are harmful to the individual affected, some changes are benign, and on rare occasion, some are helpful. It is the helpful mutations that help plants and animals adapt better to a deadly stress.

Mutations also explain speciation. The genetic code gets altered with each new favorable mutation. A favorable mutation is one that helps the species cope with its current problems. There may be a visually detectable change in the species or not. For example, bigger, more rotatable ears could help a species avoid predators. A better functioning brain would also help but might be undetectable from fossil evidence.

The combination of stresses on a species and adaptations to alleviate those stresses produce a temporarily changed species. Favorable mutations lock in those successful changes. It is the accumulation of numerous, small, locked-in changes to a species that over many generations establish a new species. We previously mentioned that horses are recognizably different from zebras and donkeys, yet all three species descended from a common ancestor. Isolation and a different environment caused the three species to become different from each other. While all three animal species are still close enough genetically to have hybrid offspring, the species have become different enough genetically that those hybrid offspring are sterile.

No matter what an individual does to change their bodies during its lifetime, the change doesn't carry over into the next generation. You can cut off the tails of mice and their offspring will still have tails. In fact, you could cut off their tails for generation after generation and the offspring will still have tails. The reason is that the genetic code for making a new mouse is in the parent mouse's DNA and that genetic blueprint is not affected by what the parent mouse does or is done to him during his lifetime. This principle applies to all creatures. However, mutations occur on rare occasions that permanently change the inheritable gene of an individual and consequently his descendants. If the mutation is harmful, the affected gene may disappear from the species over a few generations. On the other hand, a useful mutation may help the individual and his descendants survive and even thrive. The superior gene then becomes increasingly prevalent in the population over many generations. We say that it has higher frequency. This is the mechanism on how a species can permanently change over time.

18.5   WHAT DRIVES EVOLUTIONARY CHANGE?

18.5.1   ISOLATION HELPS NEW SPECIES DEVELOP

How do new species form? One effective way to create a new a species is to isolate a part of its population. In fact, one of the best natural laboratories for studying evolution is within a chain of islands. Darwin's observations in the Galapago Islands led to his theory of natural selection. He noticed that the finches were slightly different from island to island. They had adapted to the different weather conditions and different food available from island to island. Darwin was not alone in this discovery. Alfred Russel Wallace made similar observations about the animals on the islands of the Malay Archipelago. Islands are an ideal natural laboratory for studying speciation.

Why is isolation important to speciation? It is important because any modification is likely to disappear again if that modified animal breeds with individuals from the main population, whereas isolation eliminates that possibility. Moreover, species change is increasingly likely when there is isolation from the original main population for very long times. The isolation must be long enough that once united again, the returning group would not want to or be capable of mating with the old species. Darwin observed that the finches were different from each other on the different islands of the Galapagos. The plants and seeds eaten by the finches were different on the different islands. The size and shape of the finch beaks was the most altered characteristic. With the benefit of isolation, the finches of each island had over time adapted to the food and other conditions of their individual islands.

We previously discussed the chimpanzee and bonobos, who were once a single species. The Congo River then became an impassable barrier between them. After about one million years of being separated, the two apes are considered separate species. They are physically different to the trained eye, but their social behavior is the biggest difference between them. Chimps are male-dominated and aggressive, whereas bonobos are female-dominated and use sex to settle disputes rather than violence. The social differences may be due to the fact that food is more easily found south of the Congo River than north of it. Bonobos consequently live a more stress-free life than chimps.

18.5.2   WHAT ARE THE THREE MAIN DRIVERS OF EVOLUTIONARY CHANGE?

The three main drivers of evolutionary change are: (i) stresses that might kill or weaken individuals, (ii) obstacles to passing one's genes forward into the next generation, and (iii) arbitrary changes due to random mutations. The first driver is what Darwin discussed in his book on origin of the species; the power of natural selection. The second driver is what Darwin discussed in his book on sexual selection. The third driver is one of potential. Favorable mutations may be of small value to an individual or of vital importance. If the mutation has survival value to life-threatening stresses, then it gains frequency quickly. If the mutation enhances attractiveness to the opposite sex, it also gains frequency. If the mutation gives only a small advantage to survival or reproductive success, it gains frequency slowly. All three of these drivers are at work at any time. It is the different situations that determine which species will be affected and how they will change.

18.6   STARVATION AND NATURAL SELECTION

Warm-blooded animals, like us, must burn a lot of fuel (i.e., food) to maintain our internal temperature. Cold-blooded animals, like reptiles, can survive 10 times longer without food. If we don't eat, we will die. Starvation may be the most important deadly force driver in mammalian species change. The main reason that there are now thousands of mammal species or bird species where there were once only a few is their motivation to find food and eat. Species form when a group of animals exploit a new food source. Natural selection then works to optimize the evolving new species to be better and better at exploiting their ecological niche.

18.6.1   A BIRD THAT SWIMS AFTER FISH

Let me give you an example of evolution driven by the need to eat. Penguins are birds that no longer fly but now out-swim fish, catch them, and eat them. Sometime very long ago, ancestors to the penguin learned it was easier to catch and eat fish than to subsist on whatever they had been eating. Once that food source determination had been made, their bodies evolved to get better at exploiting the new feeding strategy. The more efficient fish-catching penguins displaced the less efficient ones for generation after generation. We are talking about many hundreds of generations to fully change into a fast-swimming bird. Penguins, better capable of holding their breath underwater, could catch more prey, and their offspring inherited this ability. The same is true for developing a fish-like, streamlined-body shape to help them out-swim their prey. The same is true for retaining body heat in cold water. Waterproof feathers were selected. Their wings, initially adapted to flying, became flippers. Now, they can no longer fly. Each evolutionary step of the way was driven by the need to eat and feed their young.

18.6.2   INSECT-EATING BATS

Let's do one more example by considering bats. There are many species of bats, but let's think about bats that feed on insects nocturnally. Bats do not have to compete with the birds because they found their own ecological niche. Birds can feed on insects as long as there is light to see. However, as soon as the sun has set, the hungry bats come out to feed. These bats can "see" as well with their sonar (i.e., echolocation) as birds can see with their eyes, and maybe even better. Night-time insects were a source of food that no one could harvest until the bat found a way. The bat is a remarkable animal. It screams at high frequency several times a second to generate the sound wave. His ears disconnect to protect him during each scream. They reconnect to listen to the echo with amazing sensitivity in between the screams. They can coordinate echolocation, flying and breathing all at the same time. The high demands on the body to do all three activities in synch have kept bats small.

Now bats are especially prone to starvation because of their small size. So, the driver to get good at their trade was and is a powerful motivator. Obviously, they were not always so well-adapted to nocturnal hunting. They once had eyes that saw and their sonar was undeveloped. Scientists speculate that bats began as gliding animals and learned to fly later. It has recently been proved that their echo-locating ability evolved after flight. Throughout this complex and impressive series of adaptations to become a modern bat, the threat of starvation pruned the losers and rewarded the winners at every step in the process. It has also been observed that bats will share food with a begging bat by regurgitating some of the food they have swallowed. Individual bats live perilously close to starvation on a daily basis, and this social behavior evolved because it is beneficial to the bat community.

18.7   STARVATION AND HUMAN EVOLUTION

Our closest living relative, the chimp, has a diet that is probably similar to what it was when our distant ancestors began a separate evolutionary path millions of years ago. Fruits, nuts, insects, and tender shoots were readily accessible in the tropical forests. It was a global cooling trend that led to the ice ages and the transition of tropical forests into open woodlands, and then into savannahs and deserts. We will see in the chapters ahead how hominids adapted to these climate changes with a disappearance of the traditional foods that tree-dwelling apes relied upon.

18.8   PREDATION, PARASITES, DISEASE, AND EVOLUTIONARY CHANGE

18.8.1   PREDATION

The fastest way to change a species is to apply deadly force. Consider the case of predator and prey (e.g., deer and cougars). Over countless generations, predator and prey have adapted to each other's strengths and limitations and have managed to maintain a healthy population balance. However, if either of them develops a superior trait, the other species is in danger. If the deer become much better at avoiding cougars, the cougars will starve and die. If the cougar gets more efficient at killing deer, the deer may be wiped out.

Prey animals such as mice, rabbits, and deer must have excellent senses such as hearing, sight, and smell to detect and avoid predators such as owls, hawks, snakes, coyotes, cougars, etc. Camouflage, stealth, and speed also factor into the survival of prey animals. If a prey animal is a little bit off his best game, he might become a meal. When predation goes on generation after generation, the prey animals become better and better at avoiding being eaten through the process of natural selection. However, the predators have to become better too, or they would starve to death. So, predation changes both prey and predator to evolve into better avoiders or stalkers as the case may be. In a sense, the prey animals are themselves predators. They prey on plants. The plants fight back by becoming more poisonous or more resilient at re-growing the lost part of the plant. The plant eaters in turn adapt to the changes in the plant.

18.8.2 PARASITES

Malaria is a parasitic disease that kills millions of humans every year. One evolutionary response to it is the gene that produces a sickle cell. Without the sickle cell, a person has limited resistance to the parasite. If you inherit the sickle cell gene from only one of your parents, you have good resistance to the parasite. However, if you inherit the sickle cell gene from both parents, you may likely die from anemia. The unrelenting deadly stress of the malaria parasite maintains this gruesome situation generation after generation. Half of the population benefit from the genetic anomaly, whereas one quarter pays a deadly price.

18.8.3 DISEASE

Consider the AIDS epidemic in central Africa. Now try to imagine it happening without any help or intervention from the modern world. We also assume that the Africans will not believe that the AIDS disease is spread by sexual intercourse and refuse to use condoms or limit their partnerships. Left alone to continue their careless ways, it would seem that they are all destined to die from AIDS. However, it turns out that occasionally a rare individual is born with an immunity to the disease. There is something different about their DNA that protects them from the AIDS disease. So, in our hypothetical example, almost everyone ends up dead due to the AIDS epidemic. Yet a few immune individuals survive. Those few individuals left could possibly repopulate the area with a new AIDS-immune population. In this case, their physical appearance is no different from those that died, but they are genetically superior in resistance to this particular deadly threat.

Does that scenario sound fantastic? It is documented that when the European white man first came to both North and South America, he decimated the Native American Indian populations by infecting them with diseases like measles, mumps, and whooping cough. The American Indians had no natural resistance to these diseases, whereas the Europeans had developed enough immunity to these diseases so that they didn't die from them anymore. This immunity, which developed over centuries of exposure, is in a way like our hypothetical AIDS-in-Africa story.

18.9   HOW FAST CAN AN EVOLUTIONARY CHANGE OCCUR?

18.9.1   *IGNORE WHAT HOLLYWOOD CLAIMS ABOUT MUTATIONS*

Hollywood hasn't helped with its horror movies where giant insects, rats, or whatever are created by some chemical or radiation event within a generation or two. Those movies are to entertain you, not educate you. In real life, it takes a lot longer for changes to occur. How long? It happens over many generations. So, if you want to observe evolutionary changes in a species, select a species with a fast reproduction rate. Humans have roughly 20 years between generations. Let's face it; we aren't around long enough to see many generations of live humans first hand. On the other hand, I have lived long enough to observe a noticeable increase in height of the younger generations. I was 5 ft. 10 in. tall in high school and most of the girls were shorter than me. These days, it is common for a young woman to be taller than me. Young men are even taller. They say it is due to a better diet, so it may be a reversible change rather than a permanent one. Whatever the cause, it is a significant change.

18.9.2   *THE RUSSIAN FOX STUDY*

Cats and dogs can reproduce four or five times faster than us, so you could study them over generations. Dimitry Belyaev studied silver foxes in

Russia over a 50-year period with the goal of artificially breeding for tameness. He had pondered the behavioral as well as physiological changes that wolves had undergone. He well knew that foxes are afraid of man and tend to move to the back of their cages when man comes in the room. Adult foxes might bite the handler if he attempted to touch them. In this study, the handler attempted to pet fox pups while feeding them and assign them a score for overall tameness. At sexual maturity (7–8 months), one group with only the tamest foxes was allowed to breed. There was also a control group, which was allowed to breed.

The experiment lasted for forty generations, and less than 20% of the foxes were in the experimental group each year. The results were amazing! Although the control group showed no changes, the experimental group was transformed. The specially-breed foxes were eager to hang out with humans, whimpered for attention, and licked their caretakers. The physical changes were most surprising. They developed floppy ears, short and curly tails, changed fur coloration, had different skulls and teeth, and lost their fox musky smell. So, radical changes in wild animals can evolve in as little as forty generations under the right circumstances.

18.9.3   SPEED OF SPECIES CHANGE DEPENDS ON REPRODUCTION RATE

No matter how interesting the silver fox study was, few of us are willing to devote our entire adult life to a single experiment. So mice and rats are better subjects for fast studies. They can mate at six weeks of age and have a litter three weeks later. The female can get pregnant again in a short time. If mice or rat studies are still too slow for you, consider fruit flies. These insects lay hundreds of eggs on very ripe fruit or vegetables. Within 24–30 hours, the eggs hatch into larvae (i.e., maggots). The transformation of larvae to adult fruit flies happens in under two weeks, and they are sexually active a couple days later. Fruit fly studies have made immense contributions to biological science, but I will let the reader pursue that topic on his or her own. The Internet contains a lot of info on it.

18.10   CAN A SPECIES EVOLVE BACK TO A FORMER STATE?

The probability that an evolved species could return to an earlier form is so remote that scientists believe it would never happen. One scientist, in particular, developed this concept of non-reversibility so well that it is named after him. He was a nineteenth century paleontologist named Louis Dollo. The concept that evolution is non-reversible is Dollo's Law.

Obviously, the longer the time period for evolution to alter a species and the more extensive the alteration, the more unlikely it is that a reversal is even possible. For example, the wings of birds, bats, or ancient pterosaurs were at one time arms like our own or perhaps the forelegs of a quadruped. Scientists reject the notion that nature had a long-range plan to convert their arms into wings. Evolution does not include long-term planning. Each minute change that occurred on the evolutionary path to wings was of survival value to the animal at that point in time. Of course, once the animal was capable of flying, those evolutionary changes that made it a better flyer and reaper of airborne insects were rewarded. The rewards were a longer healthier life and offspring to continue the species.

Similarly, whales and dolphins have made the great evolutionary journey from once being land mammals to becoming fully adapted sea animals. A quadruped, who traded in its limbs for fins and tail, traded nostrils for a blowhole on the top of its head, lost its fur and replaced it with blubber, became streamlined for speed in the water, and made countless other adaptations to life in water. It is ludicrous to think that whales and dolphins could revert into the land animals from which they evolved. Land is now a hostile place for whales. Occasionally, whales do get beached on land. Unfortunately, they quickly die there!

The birds that live today descended from a particular feathered-dinosaur about 100 million years ago. However, modern birds are quite different from that feathered dinosaur. The birds have gotten really good at being birds. The dinosaur only used feathers for retaining warmth. Birds have evolved stiff feathers for flight in addition to those downy feathers for warmth. The trip back to becoming the ancestor dinosaur again is not going to happen.

From a genetic viewpoint, the instructions for making the ancestor animal have been inactive for eons. Many genes used by the original animal either no longer function or they have been put to a different use. Mutational damage to the original animal's genetic code is ever more likely with passing time.

18.11   SUMMARY

In this chapter, we examined many of the factors that govern and drive species change. Natural selection rewards the fittest individuals by survival and having progeny. Natural selection culls the weakest individuals by death, and sexual selection acts to prevent them from having offspring. Species have become well-adapted to their current situations due to generations of natural selection. New stresses cause the species to change. Stresses might be disappearance of their traditional foods, climate change, new predators, and other situations that threaten their livelihood. Diversity may allow certain individuals to survive while others succumb. Over many generations, favorable mutations lock in the changes needed to overcome the threats.

We shall apply these evolutionary principles to our study of human evolution. We humans are very different from the tree-dwelling apes from which we evolved. However, the changes were incremental and important for our ancestors at the time they occurred. It was the accumulation of numerous small changes over at least five million years that account for the large differences between apes and man today.

CHAPTER 19

THE RED QUEEN EFFECT

CONTENTS

19.1 SCOPE

In Lewis Carroll's book *Through the Looking Glass*, Alice is running in pace with the Red Queen, who is a living chess piece. Alice says, "Well, in our country, you'd generally get somewhere else – if you ran very fast for a long time as we've been doing." The Red Queen responds, "A slow sort of country! Now, here, you see, it takes all the running you can do to keep in the same place. If you want to get somewhere else, you must run at least twice as fast as that!" The concept has been applied to evolution in the hypothesis that organisms must constantly adapt, evolve, and reproduce in order to survive against a world where its enemies are becoming more efficient at destroying you. In this chapter, mostly based on material from Matt Ridley's book *The Red Queen*, we will examine the Red Queen effect operating in nature and how sexual reproduction has advantages over asexual reproduction in a species survival. The Red Queen effect is as important to humans as it is to other animals. During our evolutionary history, we spend the earliest periods as a prey animal but then became a predator ourselves. However, the microorganism world may be even more important than the macro world. Parasites, bacteria, and viruses have been and still are stressors that influence our destiny.

19.2 COEVOLUTION BETWEEN COMPETING SPECIES

We have all seen films of the vast herds of wildebeests, zebras, and other grazers on the Serengeti plains of Africa. We have also seen how lions, leopards, and other predators stalk them. The predators must kill to eat at regular intervals or die of starvation. The grazers must avoid being killed in order to survive as a species. The most successful predators are able to raise and feed their offspring, whereas those who cannot find a meal before their energy runs out will never be able to reproduce. The prey animals that survive must have better hearing, sense of smell, eyesight, running ability, etc. Prey and predator are united in a coevolutionary battle for survival. Species extinction can be the price for one of them not keeping pace and thus losing this contest.

Let us consider an example of the Red Queen effect in the ongoing battle between prey and predator. The Serengeti grazers time their birthing to be all at the same time. Their newborns are especially vulnerable to predators and many will be taken, but if enough are born at the same time, some will have time to develop the required survival abilities. Again, we have all seen films of the newborn wildebeest struggling to stand upright on his wobbly legs as the mother encourages him to try harder. Lions and other predators can easily take the newborn grazers, but they cannot take them all. Enough baby grazers quickly develop the ability to run so that the species continue. Contrast that image with the helplessness of human babies for countless months. The difference is that human parents can protect that helpless baby from deadly threats, whereas the newborn wildebeest must quickly develop stability and speed if it is to survive.

We have the same Red Queen scenarios in the American West. Consider the life of deer living in the Sierra Mountains. Mountain lions are waiting for a moment of weakness to make the deer a meal. Deer with the best eyesight, hearing, and sense of smell have an advantage over the less advantaged deer in avoiding being eaten. Speed and agility are also important. Being able to outrun the cougar is a matter of life or death. However, the same arguments regarding keen senses and stamina apply equally to the cougar. If a cougar cannot bring down a deer often enough, it will starve to death. The deer and the cougar are in a never-ending arms race. Neither one of them can let the other win.

On a micro-level, the Red Queen effect operates too. There is a co-evolutionary contest between host and parasite, or host and disease germ. Malaria is one of the great parasitic killers of humans in the world. The malaria mosquito has found the human body to be a great place to lay its eggs. Humans have tried to resist the onslaught. Sickle cells give protection against malaria, which is an advantage if you could inherit the gene from only one parent. However, if you got it from both parents, you will probably die from sickle cell anemia.

19.3 SEXUAL REPRODUCTION

19.3.1 ASEXUAL REPRODUCTION

An area of interest to theoretical biologists is the debate over the relative advantages of asexual reproduction versus sexual reproduction. Asexual reproduction is the most common form of reproduction for single-cell organisms. In essence, the mother creates a female clone of herself. There is no need of a male. Asexual reproducers include archaebacterial, eubacteria, and protists. Some plants and fungi reproduce asexually as well. By contrast, most multi-cell organisms reproduce sexually.

The question becomes, "Why does sexual reproduction even exist?" It is mathematically inefficient. Two females can produce twice as many offspring asexually as a male and female can produce sexually. Computer simulation programs that match asexual versus sexual reproduction scenarios find that asexual reproduction wins every time. Actually, there is one exception. When parasites are included in the program models, sexual reproduction has an advantage.

19.3.1.1 Repelling Invaders

The Red Queen effect is involved here. The continual battle between parasite, germ, or virus versus the host is more likely to be won by the invader when considering the offspring born by asexual reproduction. The invader is much smaller than the host and may have hundreds of generations of reproduction during the lifetime of the host. Consequently, they have a new opportunity with each new generation to invent a way to break down the host's defenses. Think of your immune system as a combination lock and the invader as one who tries different number combinations in an attempt to open the lock and attack you. Each mutated invader is like a new number combination. If, as in asexual reproduction, the offspring of the host is a clone of the host, then the invaders have the same target in the host and all of his descendants.

19.3.2 A CHANGING TARGET

However, in the case of sexual reproduction, the offspring are different from either of the parents. Moreover, the hosts' offspring and their progeny are all different from each other. The invaders may defeat a few of the new host generation, but others will be immune due to their differences. Sexual reproduction provides another advantage too, that of polymorphisms. Although a fancy sounding term, polymorphisms are simply multiple versions of a gene. You have heard of dominant and recessive genes; this simply means that the genes contributed by the mother and father might not be identical, and one version is more likely to control what trait you get. With regard to the Red Queen effect and the host-invader arms race, polymorphisms add another obstacle to the invader's task.

19.4 SUMMARY

The Red Queen effect operates almost everywhere in nature. Predator and prey, invader and host, and other co-evolutionary pairs engage in an ongoing battle for survival. If one of them gains an advantage, it threatens the survival of the other species. Like the Red Queen, they must run just to stay in one place. Sexual reproduction is a less efficient system than asexual reproduction if the number of offspring is the sole criteria. However, sexual reproduction has significant advantage in protecting a host species against the invasion of parasite, germ, or virus. Instead of offering the same biological target to the invaders generation after generation, each individual offspring is different from either parent.

CHAPTER 20

EVOLUTION OF BIPEDAL APES AND HUMANS

CONTENTS

20.1   SCOPE

In Part II, we looked at who some of the prominent paleoanthropologists are, the arrival of bipedal apes in the fossil record and how they changed over their 5 to 7 million-years record, and a description of several of the individual hominin species. In this chapter, we will pick up that thread with evolutionary theory to guide us.

20.2   WHY ARE WE SO DIFFERENT FROM OUR CLOSEST RELATIVE?

How did chimps and humans, who are so close genetically, become so different in all other ways? Some scientists like Jared Diamond regard humans as a third species of chimp. Based strictly on genetic differences, that classification may be well justified, but when we consider the physical, mental, and social differences between us, it is a harder case to make. The other two species of chimp are in many ways hard to tell apart. Let us examine them next.

20.2.1   TWO SPECIES OF CHIMPS

There are two species of chimps, which you can observe in the zoo or in the wild: (i) there is the common chimp (*Pan troglodytes*); and (ii) there is the pygmy chimp or bonobos (*Pan panicus*). Note from their species name that they have the common genus *Pan*. These two species have descended from a common ancestor and have evolved into separate species. They differ from one another due to over a million years of isolation. It happens that the Congo River formed a geographical barrier between them, which they have been unable to cross. One million or so years of separation have been enough time for obvious differences to form between chimps and bonobos (see Chapter 4). Although it is true that they are socially quite different, physically they are so similar that many, if not most, people can't tell them apart. They both are fur-covered, similar-sized apes with long arms, short legs, and ape faces. Moreover, they live in trees, have the same diet, and both are equipped to swing under branches.

Now contrast that physical comparison with the appearance of either chimps or bonobos and modern humans. The difference is huge! Physically, we are very different from chimps or bonobos. We are virtually hairless, our legs are much longer than our arms, we can walk or run on our two legs, and our skull size and shapes are totally different. Chimps and bonobos are covered in fur, have much longer arms than legs, move quadripedally when on the ground, and have small brains and muggy snouts. Of course, the biggest difference between us is intellectual. We can talk, plan, and invent solutions to almost any problem that confronts us. They cannot speak or think at a level anywhere close to what we can do.

20.2.2   OUR ANCESTRY DID ALL THE CHANGING

Now here is the point. We had a common ancestor with today's chimps about five to seven million years ago. We know that from a comparison of our genomes. Fossil evidence of our bipedal ancestors confirms this time estimate. Chimps have been tree-dwelling creatures for all of their existence, and so we think that our common ancestor with chimps was very similar in appearance as today's chimps.

If the chimps haven't changed appreciably, then all the radical physical changes must have happened in our ancestry. Radical events such as this spectacular transformation tend to spark my scientific curiosity. What happened in that 5–7-million-years period to make us so different? What evolutionary forces could cause such radical change in such a short time? These are the questions that we seek to answer in this book.

20.3   WE STILL HAVE SOME CHARACTERISTICS FROM OUR APE ANCESTRY

Despite the radical physical differences between humans and chimps, there are still features about us that derive from our primate ancestry. Monkey and ape bodies are adapted to life in the trees. Stereoscopic vision was evolved to provide good depth perception. Monkeys can grasp tree limbs with all four limbs as they walk atop them. Some South American monkeys can even grasp branches with their tails. Monkeys have tails to help them balance atop the branches. Apes also can grasp branches with all four limbs, but they move through the treetops by a process called brachiation. They swing under the limbs using their long arms and grasping hands and feet. Unlike monkeys, apes have lost their tails. Evolution works to eliminate body parts that are no longer needed, and the apes' different mode of arboreal movement didn't require a tail to help maintain balance.

Humans have many of these ape characteristics: We have eyes in the front of our heads, which gives us stereoscopic vision. We don't have tails.

Our arms are loosely bound to our shoulders like apes and that allows us to swing branch-to-branch if we so desired. Who hasn't done that branch swinging when they were young? Like apes, we have grasping hands with an opposed thumb. Unlike apes, our big toes no longer oppose the other four toes in a grasping foot.

20.4   CLIMATE CHANGE DROVE OUR EVOLUTION

Climate conditions were ideal for apes during the Miocene Epoch (i.e., 23 million to 5.3 million years ago). The tropical forests were widespread, offering food, shelter, and a safe haven to their ape residents. Their numbers multiplied, new species evolved, and they spread throughout parts of Africa and Eurasia. The forest provided the fruits, nuts, etc., and other necessities of life. However, their paradise was too good to last. Global climate change was underway for the remainder of the Cenozoic Era. Global cooling over millions of years would change the landscape from extensive forests to open woodlands, savannahs, and even deserts. Moisture important to the tropical forests was disappearing as more and more water got locked up in the Arctic ice and glaciers.

As long as environmental conditions remain the same, plants and animals are under no pressure to change either. However, when temperatures get hotter or colder, and humidity get wetter or drier, plants either adapt or die, and the animals who eat those plants either adapt or die. Mind you, there were pockets of real estate that remained suitable for arboreal apes, although far fewer of them. Those particular apes did not have to change although they probably had to defend their territories against other desperate apes. It was the apes living in an altered environment who had to adapt or go extinct. These conditions of diminishing food, changed landscape, and life or death choices are what drive evolutionary change.

20.5   ADAPTING TO OPEN WOODLANDS

The distance between clumps of trees was continually getting bigger and bigger. Upright walking may have been the easiest, most efficient way of traversing those open distances. Trees were still vitally important to these bipedal apes. Trees contained the food they ate, trees were a place they could sleep in relative safety from predators, and trees were a refuge from predators. It is likely that the bipedal apes of the open woodland traveled in small groups with all eyes peeled for predators. They always knew where the closest refuge was situated. They may even have stationed observers in tall trees to watch for predators. These bipedal apes were very strong and could climb high on a tree trunk in seconds. Chimps of today can do the same and they are many times stronger than we humans are.

So certain upright bipedally walking apes living in open woodlands were the new paradigm. Food was scarcer and farther apart than it had been. They had to spend more time on the ground looking for sustenance. Those who became good at this new life style survived to pass on their genes, whereas those who could not adapt did not. Physical adaptation involved some major skeletal changes. Here are some of those changes: (i) the foramen magnum moved to allow them to balance their skulls over their vertical spines, (ii) they developed non-grasping type feet for walking, (iii) they eventually developed longer legs and shorter arms, and (iv) they developed a bowl-shaped pelvis to help support their guts. In other words, they became better and better adapted to walking and running, for which they surrendered some of their climbing ability.

20.6   ISOLATION FROM THEIR TREE-DWELLING COUSINS

Nothing is as important to new species development as being isolated from the main population. Without isolation, any new specialization that facilitated bipedalism and ground-living gained by our open woodland apes would be quickly lost if they mated with their tree-dwelling cousins. Perhaps the isolation was a product of the open woodland apes wandering farther and farther from the dense forest homelands of the tree-dwellers. The open woodland apes became ever more at home in the open woodland, whereas the tree-dwellers shunned the open woodland and became more specialized at tree-dwelling. Some scientists think the quadrupedal walking of chimpanzees, also known as knuckle-walking, was a late specialization of the dedicated tree-dwellers. It is deemed likely that the common ancestor of chimp and human was far more comfortable with upright walking and so the bipedal adaptation was not that difficult.

20.7   ADAPTIVE RADIATION OF BIPEDAL APES

Bipedal ape fossils have been found over much of Eastern and Southern Africa, and the different kinds of australopiths are anatomically different enough to where paleoanthropologists assign them different species names. Moreover, new species are continually popping up. For example, Lee Berger discovered the two-million-years-old *Australopithecus sediba* in South Africa recently. The fact that we can find so many fossils of these different australopiths suggests that they were successful species at the time and their diversity suggests they were adapting in different ways at different times and in different places. Berger believes the varied combinations of primitive and advanced features in the different species indicate a complex relationship. He suggests braided streams as a metaphor for their connectedness.

As the cooling trend continued, forests were in decline, woodlands were becoming more open, savannahs appeared, and even some areas became deserts. The exclusive tree-dwellers were declining in population as a result of the diminishing tropical forests, but the bipedal apes were increasing and covering more territory. Moreover, they were taking over a wider range of different habitats. This is the description of a species radiation. As they adapted to ever more diverse environments and different kinds of plants, they became different from one another.

20.8   INTERBREEDING

The Australopithecines (i.e., bipedal apes) existed for at least three million years, which is a long period of time. They physically changed to be better walkers by evolving the modern foot with inline toes, pelvic changes, etc. Some groups adapted to woodlands, some to savannahs, some to highlands, and some to swamps, rivers, or coastal areas and other environments. The latest thinking is that there was occasional interbreeding between these specialized hominins. This seems to be the best explanation for the strange combinations of primitive and advanced traits found in some hominin fossils.

20.9   GRACILE AND ROBUST AUSTRALOPITHECINES

Paleoanthropologists talk about gracile versus robust australopithecines. The robust evolutionary line developed larger teeth and more muscular jaws as they specialized in a diet of fibrous plants and roots. The gracile Australopithecines stayed with their age-old diet of fruits and nuts. One of the gracile lines was ancestral to humans (i.e., the Homo lineage). Which particular species that might be is controversial. Perhaps we haven't even discovered its fossils yet. Nonetheless, about two million years ago, fossils started appearing that were undoubtedly more humanlike than we had seen to date. We began to give them the genus name "Homo." We discuss them next.

20.10   ADAPTING TO NEW VEGETATION

As conditions became progressively cooler for the australopiths, their traditional foods of fruit and nuts were becoming harder and harder to find. It was no longer a problem solved by spending more time on the ground. They had to survive on whatever plants that were available. The robust australopithecines survived this period by surviving on roots and a more fibrous, hard to chew diet. We see from their fossil skulls from this period that they evolved huge molars and premolars, massive jaws and skull bones, and even a bony ridge, traversing the top of the skull, to support large chewing muscles. The change in dentition and musculature occurred over hundreds, perhaps thousands of generations. Those with stronger chewing ability survived to pass on their genes and those with lesser chewing ability did not. These robust australopithecines survived in an environment totally different from the fruit-bearing forests for which they had originally been adapted. They were successful for a long time but eventually disappeared. The concurrent Homo lineage was more successful, basing its existence on a different diet.

20.11   ADAPTING TO A MEAT-ORIENTED DIET

One group of bipedal apes adapted to a disappearing traditional food source with a different strategy; they pursued a meat diet. Most paleoanthropologists who have expounded on this topic reject the premise that early homo instantly became a hunter and killer of game animals. They think he was a scavenger first. Perhaps he watched predators eat their fill and moved in to steal some of the remaining meat and escape quickly to safety. *Homo habilis* is also known as "Handy Man" because he is associated with stone tools. Certain rocks, like flint or obsidian, break, leaving a sharp cutting edge. It would have been important to sever the meat from a carcass quickly and escape before the predator returned. *Homo erectus*, who came later, evolved to become an accomplished hunter. He is known to have tamed fire and developed better stone tools. He is believed to have cooked his meat and perhaps vegetation as well. These people were hunter-gatherers. Meat was not available whenever one wanted it. Survival depended on gathering berries, honey, insects, small animals, etc.

So we know how the robust australopithecines changed bodily to their new diet of fibrous vegetation by developing big teeth, jaws, and chewing muscles. So how did the Homo lineage physically change to excel at their hunter-gatherer life style? The most astounding change was an expanding brain size. Over two million years of evolution, their brains grew to triple that of the earliest australopithecine. This increase in head size had consequences, such as the difficulty of childbirth, limits to expanding the birth canal, and extended childcare. The female hips are limited in how wide they can get to accommodate a large baby's head before the woman can no longer walk. There was a limit to the size of the opening for childbirth. Nature found another solution to the problem. Infants were born with the brain only partially developed so that the head was small enough to pass through the birth canal. However, the baby was helpless until the brain grew bigger post-birth. Human babies imposed a prolonged burden on the mothers as a result. Large brains came with a price and therefore must have been very important to our species and their direct ancestors.

Unlike the robust australopithecines, the Homo lineage teeth got smaller and the face flatter. Having a more nutritious diet causes a reduction in the gut. Australopithecines have a ribcage that flares out to surround a big gut. Homo, by contrast, is the first hominid with a vertical ribcage. Homo adapted bodily to be an efficient hunter too. He became taller with longer legs and a new heat regulatory system. Prolonged running will raise a mammal's body temperature to the point where they have brain function problems. Hairlessness and increased sweat glands were probably evolved during *Homo erectus* time to allow for longer running.

20.12   SUMMARY

Climate change leading to cooler, drier climates caused some bipedal chimps to adapt to open woodlands and make the physical changes needed to walk effortlessly and stand upright. Savannahs and deserts replaced many open woodlands, and the bipedal apes made further adaptations. Robust australopithecines specialized on fibrous, difficult-to-chew vegetation. They eventually went extinct. Certain gracile australopithecines added meat to their diet and evolved over a two-million-years period into *Homo habilis, Homo erectus*, and finally *Homo sapiens*. The initially tree-dwelling ape transformed by this process into a long-legged, shorter armed, erect, and big-brained ground-dwelling hunter-gatherer.

PROBLEM SET FOR PART III

QUESTIONS

Q1.   In which of his books did Charles Darwin talk the most about the evolution of man?

Q2.   Darwin talked about three different kinds of selection at work in altering life forms. What were they?

Q3.   Darwin assumed that the traits of the mother and of the father of an organism would be averaged in the offspring. This assumption raised problems for superior traits being passed on. Mendel's work would have solved the problem for Darwin if he had known of it. Explain.

Q4.   How do the X and Y chromosomes differ?

Q5.   How does Richard Dawkins regard the gene's importance to our understanding evolutionary theory?

Q6.   In the Silver Fox experiment, wild foxes were changed into domesticated pets. Is this process natural selection or sexual selection?

Q7.   A movie tells the story of humans exposed to radiation, which caused them to revert back to tree-dwelling apes. How likely is that to happen?

Q8.   Explain why sexual reproduction has an advantage over asexual reproduction in a species surviving against parasites, germs, and viruses.

Q9.   Why did bipedal apes arise five million or more years ago?

Q10.   What two kinds of hominids followed the australopithecine period?

PART IV

DNA: A POWERFUL NEW TOOL

We have already acquired two tools for our human-origins-exploring toolkit. First, we learned about the power of the fossil record. We learned that the strata of the Earth hold a chronological record of life on Earth. We zeroed in on the strata that tell the tale of human evolution and saw that certain bipedal creatures appear to look more and more like us as we progress through time. The second tool in that toolbox is an understanding of how evolution works. Using that tool, we saw how a cooling planet forced many apes to adapt or die. Our ancestors learned to survive in non-tropical environments and develop a diet based in part on meat.

Now we are ready to add the third and final tool to that toolbox. That tool is DNA and the genetic code it contains. Part IV is a brief tutorial on what DNA is and why it is important. The race between scientists to understand the structure of the DNA molecule is a fascinating story in itself. However, modeling the double helix was only the beginning of the discoveries. The secret of life is that the nucleotides (i.e., the repeating units in the DNA molecule) hold the genetic code. The term "gene" had previously meant an entity that contained an inheritable trait. Now the definition is expanded to describe the gene as a sequence of nucleotides containing an instruction code for building proteins from specific amino acids. These proteins are used for a wide range of purposes by the body including embryonic development, replacing cells, regulating the body, and others.

The genes, as important as they are, comprise only a small percentage of the DNA molecule. Outside the genes lie switches, which can activate or deactivate the genes. However, the rest of the DNA molecule seems to be nonfunctional. This area is useful to us for unraveling our evolutionary history. Mutations are rare events but do alter the DNA sequence and provide a marker in time. When mutations happen in the nonfunctional parts of the DNA molecule, they do no harm to the individuals but do provide a reference point for tracking our evolutionary history.

The science of DNA discussed in Part IV may seem like a distraction from our main goal of unraveling human evolutionary history. Yet these principles will be put to good use in Part V when we apply them to resolving important questions.

CHAPTER 21

INTRODUCTION TO DNA TECHNOLOGY

CONTENTS

21.1   SCOPE

We had previously learned that traits are passed forward into the next generation of plant or animal by a unit of inheritance called a gene. Now we are going to dig deeper into what a gene is when understood from a biochemical standpoint. Genes are what make us what we are. It is the genes of a rabbit's embryo that makes it into a baby rabbit. It is the genes of a human embryo that make it develop into a human baby. To a biochemist, the term gene conjures up a string of chemical units that contain an instruction code. That code is used by the cells to build molecules called polypeptides. The polypeptides, also referred to as proteins, are used for building and repairing body parts. This all falls under the topic of DNA science because genes are the most important part of the DNA molecule.

21.2   BACKGROUND

In Chapter 16, we saw how advancements in science gradually solved the mystery of how a mating pair can pass their individual traits forward without dilution into their offspring. Charles Darwin was missing this vital inheritance information and that weakened his theory of evolution by natural selection. Gregor Mendel, a contemporary of Darwin, had discovered that superior traits were not diluted by blending of the parents' traits, but a parent's individual trait might be passed on intact. The unit of inheritance was a substance called a gene. Twentieth-century scientists delved further and discovered that a substance in the nucleus of our cells holds the instructions for making an offspring, which will have contributions from its mother and its father. Chromosomes were involved in that phenomena and it was discovered that they are made of a substance called DNA.

21.3   WHAT IS DNA?

Who hasn't heard of DNA? It is pretty difficult to find a person alive today who has never heard of it. Anyone who watches the TV programs "Law and Order" or "CSI" knows that DNA is like a fingerprint. It is unique to an individual unless he/she has an identical twin. Moreover, people who follow the news know that innocent people who had been sent to prison were able to get released years later due to DNA testing. When their DNA did not match the stored DNA from the real criminal, their convictions were overturned. If there is no match, the falsely accused have a chance to clear their name. Today's TV shows also teach us that DNA can be extracted from a person by drawing blood, swabbing their inner cheeks, or by plucking a hair from their head. DNA analysis is also used in determining heredity, propensity to get certain diseases, and other things.

   The biotechnology industry is based upon an intimate knowledge of DNA. In fact, they have learned to duplicate a key process done by our bodies. Our cells get instructions from our DNA on how to assemble a unique protein molecule from 20 possible amino acids connected in a specific order. The coded list of instructions is called a gene. The biotech industry snipped out a gene, added it into a bacteria's DNA, and put the modified bacteria to work cranking out those vital proteins. Human insulin is an example. The process is called recombinant DNA and can potentially be used to produce any natural protein produced by the body.

Scientists have even solved the problem of restoring and identifying the DNA of extinct creatures. For example, they have compared the DNA of Neanderthal fossils with the DNA of modern humans. This is remarkable considering that Neanderthals have been extinct for tens of thousands of years. DNA technology also played an important role in making this book possible. Just as fossils allow us to understand the evolutionary history of plants and animals, DNA contains similar information. If you are new to DNA science, the next few chapters are designed to review the basics. This information is important because we shall use these principles to better understand the Ape-to-Man evolutionary transition. Helping the reader understand DNA's role in paleoanthropology is the purpose of Part IV of the book.

21.4   GENES AND DNA

Genes are the units of inheritance. A single gene sometimes controls what a particular trait will be like. More often, traits are controlled by many different genes. Even then one of those genes might mutate and have a deciding influence on the trait in question. Richard Dawkins made this point in his book *The Selfish Gene*.

Parents with superior genes are likely to have offspring with superior genes. To a biochemist, the term gene conjures up a string of chemical units that contain a chemical code. DNA is composed of chemical entities that are repeated billions of times. The repeating unit is called a nucleotide, and there are four different kinds of them. How they are arranged sequentially comprises the genetic code. That code is used by the cells to build molecules called polypeptides. The polypeptides are used for building and repairing body parts. This all falls under the topic of DNA science because genes are the important part of the DNA molecule. Genes are not the only kind of entity in the DNA molecule. In fact, most of our DNA seems to have no useful function. We shall discuss this topic in detail later.

The genes in our DNA define us as human beings. The genes in a chimpanzee's DNA define it as a chimpanzee. It turns out that the DNA of chimps and humans is more than 98% the same. The difference between the DNA of humans and other apes is also close. In fact, the DNA of all mammals is closer than it would be to a reptile or bird. All life is based on DNA and all life is related to each other, some closely and some more distantly.

21.5   THE RACE

The discovery of the structure of the DNA molecule was the final step in the process of understanding the inheritance process. Moreover, it opened the door to DNA science and all the benefits that meant for medicine and other fields. So yes, it was a big deal! The race to be the first to discover the structure of DNA is one of the most exciting and important stories in the history of science and we tell that story in the next chapter. The names of James D. Watson and Francis Crick will go down in history as immortals for their work in uncovering the secret of all life on this planet. However, they were carried on the shoulders of giants who laid the foundations of DNA discovery. The following discoveries comprise the prelude to the race.

21.5.1   PRELUDE TO THE RACE

Although certain individuals are honored and remembered for their important discoveries, it is increasingly true that they are only building on the less dramatic discoveries of scientists before them. The story of the prelude to the DNA structure's race is told in Daniel J. Fairbank's book *Relics of Eden* (pp. 220 to 233). Here are a few of the contributions by pioneering scientists:

In 1869, Friedrich Miescher isolated a substance from the cells that had high phosphorus content. He called it nuclein since it was found in the cell's nucleus.

One of the great DNA pioneers was Phoebus Levene. He identified the sugar component of RNA as Ribose in 1909, and in 1929, he identified the sugar component of DNA as deoxyribose. The shorthand name DNA actually stands for **D**eoxyribo**n**ucleic **a**cid. Levene also invented the term nucleotide. You will encounter this term repeatedly in modern biological papers. The nucleotide is the repeating unit in the DNA molecule. Actual DNA molecules in plants and animals can be extremely long. They can be hundreds, thousands, millions, even billions of nucleotides long. For example, human DNA is 3.2 billion nucleotides long. Levene established that each nucleotide consisted of the sugar deoxyribose, a phosphate ester, and one of the four bases (guanine, adenine, cytosine, and thymine). The importance of these four bases to how the genetic code works in living things will be discussed in Chapter 23.

In 1944, Oswald Avery and other scientists experimentally proved that the concept that genetic information is carried on a cell protein was incorrect and that the real carrier is the DNA molecule. All of these discoveries brought biologists so close to finally understanding how living things operate and reproduce beings just like them. The remaining problem was to establish how the DNA molecule was assembled; that is, to determine its structure. When chemists talk about structure, they mean that knowing the components is not enough. They want to know how those components are arranged in a three-dimensional model of the molecule. In the case of the DNA molecule, it was both an important question and a difficult one to solve. The brilliant solution to the problem is revealed in the following chapter.

21.6 SUMMARY

Genes are the units of inheritance that are responsible for passing the individual traits of the parents to the offspring. Genes also are coded sequences of four different nucleotides and they contain instructions for building polypeptides (a.k.a. proteins) in the cells of our bodies. Genes are the most important part of our DNA, but they are not the only part. In fact, they are a small percentage of the DNA molecule. Scientists had been laying the groundwork for determining what the composition of genetic material is for many decades. The last step in the process was to understand the structure of the DNA molecule and from that knowledge understand how biological life and inheritance actually works.

CHAPTER 22

THE RACE TO DISCOVER DNA'S STRUCTURE

CONTENTS

22.1   SCOPE

In this chapter, we see how two scientists, James D. Watson and Francis Crick, working in an English laboratory, managed to beat the great American chemist Linius Pauling in the race to discover the structure of the DNA molecule. Once the double helix structure was confirmed, they solved a second problem, which is the mechanism by which DNA duplicates itself during cell division.

22.2   THE RACE TO IMMORTALITY

Human life spans are shorter than we would like. We spend a large part of our lives in school preparing for our careers and then we only have a few decades to achieve something outstanding, which will give us fame and a kind of immortality in the minds of the world community. Isaac Newton, Galileo, Charles Darwin, and other great scientists reached this pinnacle of success. Now the prize was available for the first discoverer of the chemical structure of the DNA molecule and an explanation of how it and the genes within it operated in the world of living things.

Mendel taught us that discrete units of inheritance called genes are associated with specific traits and these traits can be passed to offspring in an undiluted form. By the early twentieth century, powerful microscopes and x-ray diffraction technology allowed better equipped scientists to look at inheritance anew. They had learned that a substance (i.e., DNA) existed in the nucleus of our cells, which controlled inheritance. The next step toward unraveling the mystery of life was to identify its chemical structure. During the 1950s, two competing teams sought to solve this complex, difficult and vastly important problem.

As often happens in science when the time is ripe for an important discovery to be made, two or more individuals or teams are in a tight race to be the first to crack the problem. So it was with the discovery of the DNA structure. One team was headed by the world famous American chemist, Dr. Linius Pauling, winner of two Nobel prizes. The second team consisted of two unknown scientists in England, namely: James D. Watson and Francis Crick. For the Englishmen, deciphering the DNA structure wasn't even their main assignment. They worked on it as best they could because they saw the great importance of it. The winners of this race would not only win a Nobel Prize but are also remembered thereafter for this landmark discovery.

22.3   WATSON AND CRICK

Watson and Crick had a big disadvantage in their competition with Linus Pauling's team. They didn't have management approval and worked on the project unofficially. However, that disadvantage was offset by the fact that they had x-ray diffraction images of the DNA molecule, and Pauling's team did not. The x-rays were taken and interpreted by Rosalind Franklin. This key information enabled them to construct a 3D model of the DNA molecule and manipulate the segments until they made sense.

With much hard work and perseverance, Watson and Crick won the day and were awarded a Nobel Prize for their important contribution to science. The story behind this exciting contest is told in Watson's book, The Double Helix. Moreover, the name of his book tells us the answer. DNA, short for deoxyribonucleic acid, is a double stranded molecule, which is in a spiral configuration.

22.4   THE POLYMERIC NATURE OF THE DNA MOLECULE

The DNA molecule is a giant molecule, known as a "polymer." The word polymer literally means many mers. The "mer" part of polymer is the repeating chemical entity and the "poly" part means many. So DNA is a molecule having many repeating chemical units (mers). In the particular case of DNA, we have a name for that mer and it is "nucleotide."

The realms of the ultra-small and ultra-large worlds are hard to grasp unless you work with them routinely. A human cell is ultra-small. We have tens of trillions of them in our body. Yet the polymer length of DNA is ultra-large (about 3.2 billion nucleotides long). So at the same time, the DNA is huge, yet so tiny that it fits inside of the nucleus of a cell. This paradox is resolved when we realize that the realm of molecular dimensions is a realm of very small numbers. The diameter of atoms is measured in Angstrom units. One angstrom unit is 1×10^{-10} meters.

This may all sound esoteric, so let's scale the DNA molecule up to a size to which we can relate: think of a pearl necklace, where each of the pearls is the same size as all of the others. Now, let the pearl substitute for the typical DNA nucleotide. Let's make it a very long pearl necklace, say billions of pearls long. After all, human DNA is 3.2 billion nucleotides in length. In order to stay in portion, our pearl necklace, representing the DNA molecule, would have to be thousands of miles long. This necklace is only one strand of the DNA molecule though. Now imagine a second necklace lying alongside the first necklace. Then give the necklaces a twist so that they spiral about each other. This is something like the configuration of DNA. Figure 22.1 shows how it looks to a chemist.

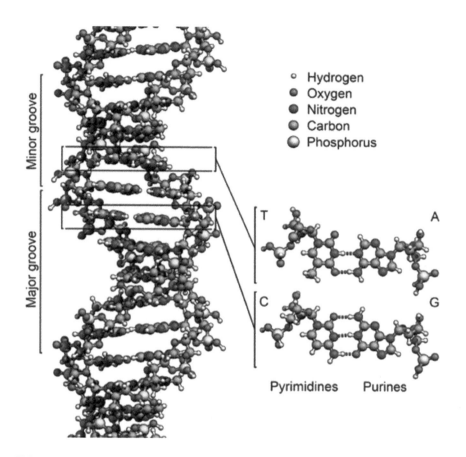

FIGURE 22.1 The DNA double helix (Courtesy of Wikimedia commons). [DNA double helix: https://upload.wikimedia.org/wikipedia/commons/thumb/4/4c/DNA_Structure%2BKey%2BLabelled.pn_NoBB.png/512px-DNA_Structure%2BKey%2BLabelled.pn_ NoBB.png]

22.4.1 COMPOSITION OF THE NUCLEOTIDES

Remember from the previous chapter that Phoebus Levene had determined the basic parts of the nucleotide. Each nucleotide has three chemical parts to it. Toward the outside of the double helix is a phosphate ester group, which links our nucleotide to the next nucleotide in the chain. Look for the locations of the phosphorus atoms in Figure 22.1. In the center of our nucleotide lies a sugar structure. Attached to the sugar is a base. The base is the variable in the DNA molecule and is very important to the differences between different DNA. The four possible bases are adenine, cytosine, guanine, and thymine; henceforth, they will be called as A, C, G, and T, respectively. You can see how they bind the two strands together in Figure 22.1. We will be discussing bases again because they are so important to how life works.

22.5 THE THREE-DIMENSIONAL REPRESENTATION OF THE DNA MOLECULE

Crick and Watson won the race with Linus Pauling to discover the 3D structure of DNA by knowing the many facts about DNA uncovered by previous scientists, having x-ray diffraction data developed and analyzed by Rosalind Franklin, and by their persistence in constructing numerous 3D models of the DNA molecule until they found the one which satisfied all of the requirements. We had used a pearl necklace to help us visualize the polymeric nature of DNA. Now we shall use a different model to illustrate the three-dimensional configuration of the DNA molecule. The 3D model is similar to a stepladder except this stepladder has a spiral shape to it. The handrails of our ladder consist of alternating sugar and phosphate ester groups. Note: Phosphate esters contain phosphate and oxygen atoms. The rungs of the ladder consist of base-pairs from our list of A, C, G, and T bases; one of the bases is attached to the sugar group on one hand rail, whereas the other base is attached to the sugar group on the opposite hand rail. There is a strong chemical attraction between the two bases, which comprise a given rung. See if you can spot the spiral stepladder shape of DNA in Figure 22.1.

22.6 HOW DNA CAN DUPLICATE ITSELF?

How does a microscopic fertilized egg become a fetus, then an infant, then a child, and finally an adult? The answer is cell division. One cell becomes two, two become four, four become eight, and soon you are looking at an organism having trillions of cells. Whatever model Crick and Watson proposed, it would have to be able to explain how this cell division process accounted for one DNA molecule changing into two identical DNA molecules. Stated differently, how is it possible for the DNA molecule to duplicate itself?

In order to visualize DNA duplicating itself, you could think of a zipper because the two strands of the DNA molecule do in fact unzip and separate under the right conditions. Instead of a zipper, however, I am sticking with our stepladder model. Remember that the two handrails of this stepladder are actually the two strands of the DNA molecule. The strands are joined by the rungs of the stepladder, which are a pair of bases from the list A, C, G, and T. We refer to these rungs as base pairs. It had already been learned that base pairs are only of two types: either the combination of a C base and a G base or the combination of an A base and a T base. Normally, the attachment between the base pairs A-T or C-G is very strong and the double-stranded molecule stays together. However, the cell can change the environment in the nucleus such that the base pair attraction is diminished and the two strands separate. This separation is essential for the DNA molecule to be able to make a copy of itself.

22.6.1   CELL DIVISION, A.K.A. MITOSIS

In the process of cell division, one of the important steps is the DNA molecule making an identical copy of itself. This process is called "mitosis" by biologists. Essentially, the two strands of the DNA molecule unzip and a new partner strand is constructed for each of the separated strands (Figure 22.2).

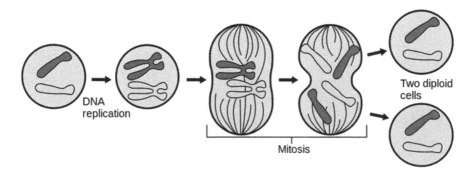

FIGURE 22.2   DNA replication (By Mysid-Vectorized in CorelDraw by Mysid from http://www.ncbi.nlm.nih.gov/About/primer/genetics_cell.html, Public Domain).

Now let's return to our stepladder analogy. Imagine that this stepladder was able to split itself into two pieces right through the center of those rungs. Let me carry this analogy a bit further. Someone has sawed all of the rungs in two and attached very powerful magnets to the sawed ends of the rungs. When the magnets are energized, the rung halves are effectively one-piece rungs again, strong and rigid. With the magnets energized, the DNA is double-stranded. However, if the magnets are turned off, the rungs split apart, the DNA molecule unzips, and we have two single-stranded halves of the giant molecule. This ability of the double-stranded DNA molecule to unzip is key to its ability to duplicate itself. In actuality, there are no electromagnets holding the rung ends together. The rungs are actually base pairs where the chemical bases having a very strong attraction to their partner base. The unzipping process is accomplished not by turning off a magnet, but by changing the acidity of the environment, which causes the bases to lose their attraction for one another.

Once unzipped, the cell containing the DNA molecule builds a new partner strand for the two divided strands. Where previously, one DNA molecule stood, two identical DNA molecules now stand. The building process for the new strands follows a simple rule: if you see a nucleotide with a C base, match it with a nucleotide having a G base. If it were a G base, match it with a C, also match a T with an A, and an A with a T.

So James D. Watson and Francis Crick discovered the chemical structure for DNA and gained immortality in the world of science as a result of deciphering the structure of the DNA molecule. They had solved a very important problem. But their next act surpasses even that one because they figured out the secret to life.

22.7 SUMMARY

The DNA polymeric strands are like pearl necklaces, where each pearl represents the repeating unit called a nucleotide. The 3D model of the DNA molecule is most similar to a spiraling stepladder, where the handrails consist of alternating sugar and phosphate ester groups. The rungs of the ladder consist of base-pairs C+G or A+T. One of the bases is attached to the sugar group on one handrail, whereas the other base is attached to the sugar group on the opposite hand rail. The strands are held together because there is a strong chemical attraction between the two bases, which comprise a given rung.

In cell division, DNA duplicates itself by unzipping to become two strands. Then new second strands for each are manufactured by the cell by following the rule that a "C" nucleotide goes with a "G" nucleotide and a "A" nucleotide goes with a "T" nucleotide. Crick and Watson thereby solved the very important question of "What is the structure of the DNA molecule?"

CHAPTER 23

DISCOVERING THE SECRET TO LIFE

CONTENTS

23.1 SCOPE

Genes are special blocks of code found in the chromosomes. Thus, the gene is seen in a new light. It is no longer simply a mysterious discrete unit of inheritance. Now we know more about what it is and how it functions. It is made of DNA material and it is an instruction code for building a polypeptide molecule from a string of amino acids. One or more polypeptides combines to make a protein. Sometimes we shorten this to saying that genes provide instructions for building proteins. In this chapter, we shall see how this process is carried out.

23.2 WHAT IS A PROTEIN?

Proteins do most of the work in cells and are required for the structure, function, and regulation of the body's tissues and organs. They serve as antibodies, enzymes, hormones, body structure, and transporting agents.

Proteins are also polymeric molecules, although they are much smaller in length than our DNA molecules. They might be composed of dozens, hundreds, or even thousands of mers (i.e., repeating units). Proteins can be infinitely variable because each mer can be selected from a selection of 20 different configurations. Proteins are assembled from a special kind of chemical known as an amino acid. Like the name implies, there are two reactive groups as part of each amino acid, namely: an amine group and an acid group. These reactive groups would love to combine with each other to form a peptide linkage, but their placement on the amino acid molecules makes that impossible. However, they could combine with a neighboring amino acid.

23.3   GENES AND PROTEIN PRODUCTION

One of the big jobs that DNA performs is to provide manufacturing instructions to build new body tissue. The instruction code is called a gene and the end product is called a polypeptide. One or more polypeptides combine to form a protein. Crick and Watson, after cracking the 3D structure of DNA, blazed ahead to understand what the gene is, what it does, and how it does it. The gene is a very specific string of the letters A, T, G, and C within the DNA molecule. The protein manufacturing sites in your body read these genes with a "start here" and "end here" code to them. Humans have about 21,000 genes residing in their DNA. The gene is an instruction code, which allows a particular protein to be manufactured by the cell.

23.4   RNA AND PROTEIN PRODUCTION

RNA is short for the chemical name "ribonucleic acid." RNA is similar in many ways to DNA except it is single-stranded, not double-stranded, and RNA is usually a much smaller molecule than DNA. When I think of RNA, I think of ants busily scurrying about an anthill or perhaps mobile robots delivering car components and assembling them on new automobiles moving along an assembly line. They are lifelike and busy making things happen. The difference is that RNA molecules are infinitely smaller than robots or ants and their activities result in the manufacture of those biochemical entities that rebuild our bodies that keep us from getting sick, that juice up our bodies for fight or flight action, and all the other functions necessary for survival. Those biochemical entities are proteins and they come in an infinite arrangement of their building blocks.

Let us consider two types of RNA: messenger RNA (mRNA) and transfer RNA (tRNA). Messenger RNA is active in a process called transcription, whereas transfer RNA is active in delivering those protein building blocks (i.e., amino acids) to the protein assembly site (i.e., a ribosome). The ribosome is a complex molecular machine found outside the nucleus of cells. Transcription is the first step in the protein production process, so let us discuss it next.

23.5   TRANSCRIPTION

Simply stated, transcription is the process of obtaining a copy of the instruction code in a particular gene, which is inside the cell nucleus, removing the copy from the nucleus, and attaching it to a ribosome, which is outside of the nucleus. The copy of the gene's instruction code is actually messenger RNA (mRNA). Remember that DNA is double-stranded, whereas RNA is single-stranded. So, an agent for this process must separate the two strands of the DNA molecule and build a single-stranded duplicate of one of the strands over the length of the gene. The agent able to do this job is called "RNA polymerase." The chemical composition of mRNA is slightly different from the DNA strand, which it is copying. Those differences are not important to our discussion here, and the genetic code is not altered. If you want to dig deeper into transcription, search Wikipedia using the search word "transcription." The mRNA thus generated leaves the nucleus of the cell and attaches itself to a ribosome

23.6   PROTEIN PRODUCTION

Polypeptides are produced by stringing amino acids together and inducing them to react with one another. The amino acids contain two kinds of reactive groups in their molecule: (i) an organic acid, and (ii) an amine. Due to a restricted geometry, the two groups on the amino acid molecule cannot react with each other, but they can react with an adjacent amino acid. Thus, the acid group of one amino acid can chemically react with the amine group of an adjacent amino acid forming a permanent linkage. Polypeptides result from these linkages when many amino acids are arranged in a row. The amino acids can be induced to all link together in that sequence, and such is the case in building a polypeptide molecule. One or more polypeptides are then used to assemble a protein.

These reactive groups on the amino acid molecules are important because we can induce one amino acid to permanently attach itself to the next amino acid by forming a chemical link. That link has a chemical name and it is called a peptide. So if we unite a long string of amino acids, together we have a polypeptide. Proteins are made of polypeptides.

Living things all use the same 20 different amino acids to build proteins. The variety of amino acids makes it possible to customize proteins for specific missions. The ability to string these different amino acids in any arrangement increases the variety of proteins possible. Wait, you might say, how do you get specific amino acids to line up in a specified order? Enter tRNA, whose job it is to find a specific amino acid from the list of 20 and haul it over to the protein construction site. But how does the tRNA know which amino acid to find? The answer is codons, and we consider them in the next section.

23.6.1 CODONS

James Watson developed an ongoing relationship in the mid-1950s with the famous physicist George Gamow, who was intrigued by the progress in genetic science. It was Gamow who first suggested that amino acids could be identified by a three-nucleotide sequence. The three-nucleotide codon was needed to be able to uniquely identify all of the 20 amino acids. There are 64 ways of arranging a three-nucleotide sequence, but only 16 ways of arranging a two-nucleotide sequence, and only four different arrangements of a single nucleotide. Consequently, the three-nucleotide codon can specify any one of the 20 amino acids and that is exactly what nature does when it builds a protein by assembling a specific sequence of amino acids. The sequence of codons is found on the mRNA molecule. It should be mentioned that during the transcription process, uracil is substituted by thymine on the mRNA molecule. So the DNA letters (T, A, G, and C) become (U, A, G and C) on the mRNA molecule. Table 23.1 shows a sample of six of the 20 amino acids and which codons identify them.

The cell builds proteins or their component, the polypeptide, by assembling specific amino acids in a specific arrangement and then chemically uniting them into a polymer molecule. Subsequently, this polypeptide molecule folds up into a biologically useful shape. As an example, hemoglobin, the protein in your blood carries oxygen to your cells and transports carbon dioxide out. It is actually comprised of four particular polypeptides folded into a specific shape, which surround an atom of iron.

23.6.2 PROTEIN FOLDING

The numerous different proteins produced in our cells do a myriad of different jobs within our bodies. Usually, the shape of the protein is highly important to its function. What started out as a specific string of different amino acids became a three-dimensional polypeptide molecule having a unique shape. Out of numerous possible conformations the protein molecule might assume, it folds itself into that single unique 3-D shape that makes it capable of doing its function. As we humans attempt to duplicate nature and manufacture some of the proteins useful in medicine, the protein-folding phenomenon has been hard to understand and duplicate for commercial purposes.

TABLE 23.1   Some Amino Acids and Their Identifying Codons

Amino Acid	Description	Codons
Phenylalanine	Nonpolar	UUU or UUC
Histidine	Basic	CAU or CAC
Alanine	Nonpolar	GCU, GCC, GCA, or GCG
Glutamine	Polar	CAA or CAG
Tryptophan	Nonpolar	UGG
Glutamic Acid	Acid	GAA or GAG

Source: https://en.wikipedia.org/wiki/Genetic_code.

23.7 SUMMARY

The gene can now also be visualized as an instruction code for building a protein. Proteins are a combination of one or more polypeptide molecules. Polypeptides are composed of a string of amino acids. The process first involves making a copy of the gene. Messenger RNA and RNA polymerase are used to accomplish this task. The duplicated gene exits the nucleus of the cell and attaches to a ribosome outside the nucleus. The genetic code on the gene is read three letters at a time (i.e., codons). These codons define particular amino acids, which are brought to the ribosome by transfer RNA. A polymerization reaction unites the amino acids into a single molecule called a polypeptide.

CHAPTER 24

MUTATIONS AND JUNK DNA

CONTENTS

24.1 SCOPE

The previous three chapters discussed how the science community methodically isolated DNA as the key to inheritance over multiple decades, established its composition and three-dimensional structure, determined that base pairs on the nucleotides comprised an instruction code for making proteins, and discovered that three-letter codons were used to line up amino acids to assemble the building blocks of the protein. In this chapter, we will discuss what a mutation is, how it modifies DNA, and what that might mean to the individual. We will mention that switches exist in our DNA that can activate or deactivate a gene. Finally, we will discuss junk DNA, how it evolves, and how it contains valuable information about our prehistoric past.

24.2 MUTATIONS

Mutations are what allow the inheritable traits of a species to permanently change. Inheritable traits may include: height, hair color, arm length, color vision, etc. Now, some mutations occur in the inactive part of our DNA and have little to no effect. However, mutations in our genes can have a significant effect. Specifically, what effective mutations do is alter the genes in our DNA. Genes are essentially coded instructions on how to make a particular protein in our cells. The instruction code is made up of a string of nucleotides, each representing the base designated by the letters, T, G, A, and C. Remember that the base pairs are always composed of a "T" with an "A" or a "C" with a "G." It is at the level of the base pair that a mutation occurs. One kind of mutation is *substitution.* For example, a T-A base pair is substituted for a C-G base pair. A second kind of mutation is *insertion.* A new base pair is inserted into the string of nucleotides. The third kind of mutation is *deletion.* Here, a base pair disappears from the string of nucleotides.

As a cell copies its DNA before dividing, an error occurs once on average for every 120,000 nucleotides. Genes only comprise a small percentage of our DNA, but when a mutation occurs within a gene, that cell may no longer be able to make the same protein because the code may no longer be specifying for the required amino acids. Depending on what function that protein performed, the health of the individual may suffer. Mutations might also occur in the switches that activate or deactivate those genes. However, mutations are not always bad. Sometimes a mutation has a neutral effect. For example, more than one particular codon can specify a given amino acid. If the mutated codon still specifies the same amino acid, then the mutation is neutral in its effect. Mutations can even be positive. If the altered gene yields a superior protein for the task it performs, then the mutation is a good mutation. Different versions of the same gene are known as alleles. Alleles account for the differences between individuals in a species. In humans, height, hair color, eye color, skin color, resistance or susceptibility to disease are variations all due to alleles.

24.3 SWITCHES AND REGULATORY GENES

We know that the part of our DNA consisting of genes is useful to us in providing instructions for making proteins. However, that is only a small percentage of our DNA. There are also segments of DNA that act as switches to activate or deactivate those genes. These DNA sequences are generally short and usually under 50 bases long.

Regulatory genes are genes that control the expression of one or more other genes. They may use proteins to do this or use RNA. Genes being affected by the regulatory gene may be either repressed or activated. Repressor proteins work by preventing RNA polymerase from transcribing RNA. Activator proteins bind to a site near a gene and cause an increase in transcription.

24.4   JUNK DNA

Most of our DNA seems to have no usefulness and that inactive portion of it is often referred to as "junk DNA." It is often said to be the non-coding portion of our DNA. Junk DNA appears to have no useful purpose, but more is being learned all the time. The DNA sequences between genes are non-coding, but sometimes junk DNA inserts itself inside a gene. These segments are called introns.

There was some solace in the thought that junk DNA is at least chemically locked in place and benign, but in 1950, a researcher reported that some pieces of DNA actually move around and, depending on where it transplants itself, could do serious damage to the individual and possibly his descendants.

24.4.1   TRANSPOSABLE DNA ELEMENTS

Barbara McClintock was a pioneer in DNA science. She upset the scientific establishment with her discovery that small DNA segments were moving from place to place in the genome of corn kernels. These "transposons" exist not only in corn but in the DNA of all living things, including humans. There are two kinds of transposable elements: *transposons* and *retroelements*. Transposons move within the genome by removing themselves from their current location and inserting themselves in another location. Retroelements move within the genome by making a copy of themselves and inserting that copy in another location. Obviously, retroelements have a greater chance to multiply and occupy a significant part of a genome. Daniel Fairbanks, in his book *Relics of Eden*, reported that in the case of the human genome, transposons are 2.8%, whereas retroelements are 42.8% of the genome. He further tells us that most of our millions of retroelements have quit reproducing themselves, although a few are still active.

It is believed that retroelements became inserted into our genomes through a special type of virus. The so-called retrovirus consists of RNA within a protective protein coating. Fairbanks tells us that these retrovirus infections happened tens of millions of years ago.

24.4.2   PSEUDOGENES

Pseudogenes look like genes but can no longer build a protein. They are genomic fossils that have become nonfunctional due to mutations. For example, the mutation may interfere with transcription of the gene or with translation by introducing a stop codon too early in the process. While most pseudogenes remain permanently nonfunctional, in rare cases a further mutation may make it functional again.

24.5   OUR PREHISTORIC PAST

Our junk DNA actually contains a treasure trove of information about our prehistoric past and genetic relationship with other species. Most, if not all, mutations that occur within our junk DNA are neutral. So no harm comes to those individuals from the mutation, and a genetic marker is left in their DNA. That marker will pass on to all subsequent offspring but will be missing from those ancestors living prior to the mutation. Consider our relationship to the other apes (e.g., chimpanzees, bonobos, gorillas, and orangutans). The numerous genetic markers in the DNA of these living apes along with DNA from *Homo sapiens* allow us to establish a chronological order to the markers, and from that information, we can see which species are most closely related. The same principle can be applied to humans from around the globe. Human migrations out of Africa and into Europe, Asia, Australia, and the Americas can be worked out.

24.6 SUMMARY

Mutations may now be defined in terms of nucleotides: One kind of mutation is *substitution*. For example, a T-A base pair is substituted for a C-G base pair. A second kind of mutation is *insertion*. A new base pair is inserted into the string of nucleotides. The third kind of mutation is *deletion*. Here, a base pair disappears from the string of nucleotides.

Genes can be turned off and on by the body. Switches reside near the genes and are activated by chemicals attaching to them. Regulatory genes are able to activate or retard other genes. Most of our DNA seems to be inactive and is referred to as junk DNA. Transposable elements such as *transposons* and *retroelements* have been discovered to move around in our genomes. Retroelements are capable of making copies of themselves and embedding them throughout our genomes.

Finally, neutral mutations in the DNA of living individuals provide a historic record of species and speciation. This important tool allows us to explore the past in a way not possible by other means.

PROBLEM SET FOR PART IV

QUESTIONS

Q1.   We know that genes are the basic unit of inheritance, but what is a gene physically and where does it reside?

Q2.   During the process of cell division, the DNA molecule in the nucleus makes a copy of itself during cell division. How does that happen?

Q3.   If the DNA molecule resembles a spiraling stepladder, which part of the stepladder do the base-pairs represent?

Q4.   A particular type of RNA delivers the correct amino acid to the ribosome when it is building a protein molecule. What is the name of that RNA type and how does it know which amino acids to fetch?

Q5.   Describe the process called "transcription."

Q6.   Why is a codon a three-letter sequence instead of a two-letter sequence?

Q7.   Our genes are inherited as pairs. One gene comes from our mother and the other from our father. It is possible that both inherited genes are different from each other. What do scientists call such genes?

Q8.   A mutation caused a substitution in the nucleotide of a gene. Yet, the gene still produced the identical protein. How can this be?

Q9.   Our DNA is mostly so-called junk DNA. It is not important to our survival. Retroelements account for much of this junk DNA. Explain.

Q10.   Neutral mutations may serve as markers in our DNA. These markers can help us unlock the prehistoric past. Explain.

PART V

DNA APPLIED TO PALEOANTHROPOLOGY

In this section of the book, we are going to resolve some important questions about our evolutionary history. Without the power of DNA science, it is doubtful that we would ever answer these questions. One of those questions regards our descent from the apes. To which of the existing apes are we most closely related? Chapter 26 walks us through some fundamental concepts relevant to Part V. Chapter 27 walks us through the scientific process that finally resolved this question of our ape ancestor. It wasn't an easy problem to solve. The impasse was known as the hominoid trichotomy. The solution involved an appreciation of the gene tree-species tree relationship.

Our species, *Homo sapiens*, is the only existing species of humans left on the Earth and we are everywhere. Two main theories have competed to explain our presence throughout the world, the multiregional theory and the Out-of-Africa theory. The first theory contends that we evolved regionally from more primitive people living in the areas. The latter theory contends that a wave of modern humans replaced all primitive people on the Earth. In Chapter 28, we show how DNA technology resolved this controversial debate to the satisfaction of most anthropologists.

The Neanderthals went extinct about 30,000–40,000 years ago. They had Europe and the Middle East to themselves for tens of thousands of years. The migration of modern humans into the area might have been the reason for their demise. The seemingly unanswerable question is "Did interbreeding occur between these different kinds of humans and did the offspring survive and contribute their genes to humanity? Chapter 29 tells the story of how one extraordinary scientist painstakingly solved the problem of sequencing ancient DNA, sequenced the DNA obtained from Neanderthal bones, and answered the question of interbreeding.

New species of humans have been declared after examining enough fossil bones and teeth to establish differences with other established species and prove they had found something new. How then could a new species be established from a single small bone from a little finger? In Chapter 30, we learn about the existence of the Denisovans. This species was established not from comparisons between skeletal parts but from comparisons between nuclear DNA.

CHAPTER 25

DNA SCIENCE APPLIED TO HUMAN ORIGINS

CONTENTS

25.1   SCOPE

This chapter helps us transition from the basic facts about DNA covered in the previous section to the specific DNA science used to solve problems in paleoanthropology. We can obtain DNA samples from people or animals living today and from people and animals long dead. The latter capability has been extended backwards timewise due to the innovative efforts of dedicated scientists. We have two kinds of DNA in our cells: nuclear DNA and mitochondrial DNA. The latter type is a much smaller and less complicated material than the former and it is easier to obtain. Earlier work relied more on mitochondrial DNA for these reasons. However, more information can come from nuclear DNA than mitochondrial DNA, and great advancements in performing those analyses have occurred.

25.2   ALL LIFE IS RELATED

25.2.1   AN IMPORTANT FUNDAMENTAL CONCEPT

All living species have a common ancestor with each other. Darwin grasped this important concept back in the nineteenth century, and since then discovery after discovery has confirmed it. Moreover, all living things use DNA to pass their traits forward to the next generation and to maintain their own bodies. We talk about any two living species having a common ancestor at some point in time since life first appeared. This kind of talk is based on an underlying premise that all life today is interrelated through a common ancestor in the distant past. All life on our planet is based on DNA. This fact persuades us that all existing and past life descended from a single common ancestor, also based on DNA. We have the advantage of exhaustive testing to bring us to this conclusion.

DNA science can now compare the DNA of two species and closely estimate how long ago they had a common ancestor. Before this capability developed, paleontologists relied on comparisons of fossil bones and the age of sediments to estimate the time two species separated. Due to the rarity of finding a desired fossil and the difficulty of dating sediments, the DNA analysis is a welcome tool.

25.2.2   FURTHER PROOF THAT LIFE IS RELATED

A basic building block of living structure has not changed in 400 million years. Consider the following facts, which add further evidence to the conclusion: Telemeres are sequences of DNA code found at the ends of chromosomes. They serve to prevent the decay of the chromosome during cell division. The telomere is analogous to the blank section of magnetic tape on the leader, which protects the music or film when the tape begins to scroll. The following discoveries are reported in *Scientific American* Aug. 1991, in the article "The Human Telemere." The author of the article, Robert Moyzis, operated out of Los Alamos National Laboratory. He and his team succeeded in identifying the repeating DNA code sequence "TTAGGG" in a series of experiments.

But more significantly, he proved that that same sequence also exists in fish, amphibians, reptiles, birds, and mammals. A basic building block of living structure has not changed in 400 million years, when these vertebrates all had a common ancestor. In another experiment, the human telemere was substituted for the telemere in yeast and the yeast thrived. The common ancestor for humans and yeast, existed approximately a billion years ago. We are getting very close to the age of the common ancestor of all life at one billion years.

25.3 HOW CLOSELY RELATED ARE ANY TWO SPECIES?

25.3.1 POWERFUL NEW TOOLS

Prior to the mid-twentieth century, scientists used to rely mainly on physical comparisons to determine how closely related two species might be. For example, a horse and a zebra are physically very similar. One marked difference is the zebra stripes versus the horses' lack of stripes. Otherwise, the animals are very similar. In fact, they can mate and have hybrid offspring known as "zebroids." These hybrids often exhibit both striped areas and uniform areas on their bodies. Obviously, these animals once had a common ancestor, but we previously had no way to determine when that was unless we were lucky enough to have the fossil evidence. When we are dealing with the fossils of animals long extinct, determining kinship is a much harder task than comparing two living species.

DNA technology has added a powerful tool to our capabilities. Now the DNA of the two species can be examined and the time at which they had a common ancestor can be closely estimated. We can also determine how closely related any two species are with each other. The term "genetic distance" is used in this regard. Here, we are getting DNA samples from living individuals and comparing them. It is also possible to sequence the DNA of extinct animals. Of course, there is a limit to how far back we can go. DNA decomposes over time, and foreign DNA can contaminate samples. Even so, new techniques have been found to sequence ancient DNA, which is tens of thousands of years old.

25.3.2 MITOCHONDRIAL DNA

It turns out that animals have more than one kind of DNA in their bodies. There is the DNA that resides in the nuclei of our cells (i.e., nuclear DNA), but there is a smaller and different DNA molecule that is inherited from our mothers alone and resides outside the cell's nucleus. The latter kind is called mitochondrial DNA or mtDNA. We can compare the mtDNA from two species and even estimate when they separated from a common ancestor. Mutations that have occurred in mtDNA are used as markers. Mutations common to both species represent the time when they were one species, whereas mutations, which developed after that common-ancestor-period, would be different for the two species. Scientists have calibrated the time scale at which mutations occur and, using that knowledge, estimate when the common ancestor lived. One other benefit of using mtDNA to compare species or establish a chronology is that it is exclusive to the female inheritance lineage. When sperm and egg unite to form an embryo, the egg is rich in the mother's mtDNA, whereas the sperm has a negligible amount of mtDNA.

25.3.3 NUCLEAR DNA

Nuclear DNA is a vastly larger molecule than mtDNA and is unique to the individual because it is composed of a 50/50 mixture of each parent's DNA. The difference in DNA between any two individual humans is about 0.1%. Human nuclear DNA is about 3.2 billion nucleotides in length. The vast size of the nuclear DNA molecule makes sequencing time-consuming and expensive. However, testing can be done with a particular small segment of the nuclear DNA, which can serve to trace the male inheritance lineage in the same manner that was used with mtDNA. That segment is the Y-chromosome. It happens to be the smallest of our chromosomes. If you have an X and a Y chromosome, you are a male. If you have two X chromosomes, you are a female. Consequently, testing of the Y-chromosome only involves lines of male descent. We shall see in the future chapters how the Y-chromosome revealed the journey of *Homo sapiens* out of Africa and throughout the remaining world.

25.3.4   THE GENOME AND BIOINFORMATICS

Autosomal DNA is all of the non-sex chromosomes. Testing can be conducted on defined segments of the autosomal DNA to simplify the task. However, more and more interest is developing in testing the entire DNA molecule due to the superior informational results that it yields. In fact, since the human genome project, ambitious studies of the complete genome have been undertaken. The 1000 human genome study (conducted between 2008 and 2015) compares individuals from all over the globe. A wealth of information has and is coming from this study. An entirely new science is developing to analyze DNA sequences called bioinformatics. It combines computer science, statistics, mathematics, and engineering to analyze and interpret biological data. For example, it uses computer programing to glean information from human genomes of the world. Bioinformatics has become an umbrella term for using computer programs to identify genes, single nucleotide polymorphisms (SNPs), microsatellites, and other points of interest in the genome. Although medical science is the big user of this technology, paleoanthropology is a beneficiary of the technology too.

25.4   DECOMPOSITION OF DNA

When a plant or animal dies, its DNA begins to decompose along with the rest of the body. If you saw the fictional movie based on Michael Creichton's book *Jurassic Park*, you may recall that the scientists extracted dinosaur DNA from the bodies of mosquitoes, who supposedly drank their blood. The mosquitoes, in turn, were preserved by entrapment in tree sap. The tree sap fossilized as amber and preserved the mosquitoes from the Jurassic Period until the present. The amber was mined from geological strata many millions of years old. I am impressed by the author's keen mind and imagination. However, we now know that DNA decomposes faster than that, and Creichton's thesis doesn't work. What a pity!

Some researchers have reported finding dinosaur DNA, but careful retesting showed these findings to be erroneous. We now know that DNA does not survive for millions of years. It will decompose long before then.

The conditions that are best for the preservation of DNA are cold, dry, and oxygen-free storage conditions. Mammoths, who were quick-frozen tens of thousands of years ago in the Arctic, are ideal subjects. Today, the testing of ancient DNA has developed into an exacting technology. Keys to success include: finding suitable fossils, keeping them sealed, and avoiding contamination from the omnipresent unwanted DNA.

25.5 SUMMARY

The concept that all species descend from a common ancestor and that all species have a common ancestor between them is very powerful because we can now use DNA analysis to determine that separation as a function of elapsed time. We have two kinds of DNA in our cells: nuclear DNA and mitochondrial DNA. The latter type is a much smaller and less complicated material than the former and it is easier to obtain. Earlier work relied more on mitochondrial DNA for these reasons. However, more information can come from nuclear DNA than mitochondrial DNA, and great advancements in performing those analyses have occurred.

We can obtain DNA samples from people or animals living today and from people and animals long dead. The latter capability has been extended backwards timewise due to the innovative efforts of dedicated scientists.

CHAPTER 26

TRACING OUR APE HERITAGE

CONTENTS

26.1   SCOPE

This chapter addresses a simple question, namely: of the existing apes, which species is closest genetically to humans? The problem remained unresolved for a period, where tracing the evolutionary history of different genes led to different results. Through advances in genetic science, the problem was resolved and a new understanding of "gene tree versus species tree" relationships emerged.

26.2   WHERE DO WE BELONG IN THE ANIMAL KINGDOM?

Some people don't believe that we humans are animals, but from a scientific viewpoint there is no doubt about it. We are in the animal kingdom. We are of the phylum Chordata, which means we are the group of animals having a spine. Moreover, we are vertebrates. We are in the class Mammalia, meaning we are a mammal. Like other mammals, we have body hair to retain heat, and we suckle our young. We are of the order Primates and the suborder Anthropoidea. So we are in a group that had adapted to living in trees. Apes, unlike monkeys, are tailless and are able to swing from branch to branch due to the flexibility in their arms and the gripping ability of their fingers. We are in the family Hominidae, which connects us with the other hominins like the fossil men described in Part II of this book. Our genus is Homo and species is sapiens. We are *Homo sapiens*.

26.3   WE ARE CLOSEST TO THE GREAT APES

Charles Darwin, who lived in the nineteenth century, didn't have a small fraction of the scientific information that we have at our fingertips today, yet he was astoundingly insightful when he wrote in *The Descent of Man* that "these two species [gorillas and chimpanzees] are now man's greatest allies." In other words, we are most like the chimpanzee and gorilla. Using technologies that were unimagined in Darwin's day, we have confirmed his suspicion. By comparing the DNA in the genomes of humans, chimps, and gorillas, we have confirmed that these three species are more closely related than any of them are to the Asian apes, orangutans, and gibbons.

26.4   THE HOMINOID TRICHOTOMY

During the last quarter of the twentieth century, scientists could not clearly establish the genetic relationship of humans, chimpanzees, and gorillas. The scientists wondered, were the chimps and gorillas genetically closer to each other than they were to man? Or were the chimps and humans closer to each other than either was to the gorilla? Or were the gorilla and humans closer to each other than either was to the chimp? Different experiments yielded different answers. It seemed to be impossible to decisively prove any of these three hypotheses. This intractable problem became known as the hominoid trichotomy, where the word "hominoid" refers to the super-family Hominoidea, which includes humans, orangutans, gorillas, and chimpanzees.

26.5   DNA-DNA HYBRIDIZATION

From this point forward in this chapter, I will be summarizing the facts and discussions presented in Chapter 2 of the book *Ancestors in Our Genome* by Eugene E. Harris. If this topic interests you or if you want a more thorough understanding of it, then this book is for you.

In one kind of experiment, known as DNA-DNA hybridization, entire genomes from two species are unzipped, mixed together, and allowed to recombine, and then examined to determine how well the DNA strands stick together. The better the binding is, the closer the species are genetically. The result was that human-chimp binding strength was greater than either human-gorilla or chimp-gorilla binding strength. However, this experiment did not end the debate. Critics were not convinced by these data, and some reinterpreted the data and found no clear conclusions.

26.6   GENE TREE EXPERIMENTS

The Human Genome Project determined that humans have about 21,000 genes in their genome, which is fewer than had been expected. We have since learned that primates have about the same number of genes and that there is roughly a one-to-one correlation of inherited genes among the primate species. These common genes are called homologous genes. So, our human genes once existed in the common ancestor of living hominoids, and the mutations in a gene of a living hominoid can provide information about its evolutionary history.

Remember that genes are instruction codes for building proteins and consist of a sequence of bases (T, A, G, or C). When the same gene exists in humans and the great apes, there may be an opportunity to establish the time of separation of the different species by comparing the code sequence of the gene in each of the species. Neutral mutations that occurred during the time the species that still one should be identical. On the other hand, mutations that occurred after species split apart would likely be different. And so, if we are comparing three or four species (e.g., humans, gorillas, orangutans, and chimps), then it should be possible to construct a gene tree where each species branches off at its determined exit point. Remember that the resulting tree has only been constructed for a single gene out of 21,000 possible genes. It may or may not represent the true species tree. With these principles in mind, let's examine the results from several different homologous genes.

26.7   COMPARISONS OF THE ALPHA-1,3-GT GENE

The alpha-1,3-GT gene had long ago become inactive in monkeys and apes and had acquired numerous mutations, which made it valuable for detecting differences in the genes of the hominoid species. Mutations that are identical in two species were most likely there before those species branched apart, whereas differing mutations most likely occurred after they branched off. This gene consists of 371 bases, and for most of those bases, the great ape genes were identical. Fortunately for the experiment, 21 bases were different.

Analysis of the data showed that the orangutan had branched off from the other hominoids (human, chimp, and gorilla) at the earliest point in time. It is farthest from humans and therefore, the orangutan need no longer be considered. This orangutan elimination reduced the number of relevant mutation sites to twelve, and only two of them were useful in determining evolutionary relationships. Yet this number of bases was enough to do an analysis and conclude that the most likely interpretation of the data has the gorilla branching off from the chimp-human line first, leaving the chimp and human as closest in kind of the three. The conclusion is that we are most closely related to chimps.

26.8   THE CONTROVERSY CONTINUES

However, the alpha-1,3-GT gene study did not end the controversy. The following table summarizes the findings of some other studies using different genes.

As Table 26.1 indicates, studying yet another gene is not going to resolve the hominoid trichotomy. Moreover, something strange is going on. How can different genes tell different stories about species relationships? What is going on here?

26.9 THE GENE TREE-SPECIES TREE PROBLEM

A species tree shows the evolutionary relationship between species. The point where species branch off represents the point in time where genes were no longer shared. Using the results from Table 26.1, three different species trees might be drawn for the gorilla, chimp, and human relationship.

- Tree 1. Gorilla branching off first, chimp and human branching off last.
- Tree 2. Chimp branching off first, gorilla and human branching off last.
- Tree 3. Human branching off first, chimp and gorilla branching off last.

Gene trees are different from species trees, in that the gene tree only indicates the evolutionary history of one particular homologous gene. As we saw from Table 26.1, Trees 1, 2, or 3 might be the relevant gene tree depending upon which gene we had investigated. The process of sexual reproduction is responsible for this multiple gene tree phenomenon. Sometimes during meiosis, regions of a chromosome break off and attach to a different chromosome (a.k.a., crossing over). The end result is that some gene trees are different

TABLE 26.1 Results of Gene Studies and the Hominoid Trichotomy

Date	Gene Studied	Closest-Related Pair
1989	Involucrin gene	Chimp and gorilla
1989	Immunoglobulin C alpha 1 gene	Gorilla and human
1991	CoII mitochondrial gene	Chimp and human
1992	Beta globin cluster of six genes	Chimp and human
1993	Protamine 1 gene	Gorilla and human
1993	Protamine 1 gene	Chimp and gorilla

than the species tree. That gene tree might dominate in the species population if it gave the individuals of that time a natural selection advantage.

26.10   FINAL RESOLUTION TO THE HOMINOID TRICHOTOMY

If you are interested in the details of how the trichotomy was resolved, you should read *Ancestors in Our Genome* (pp. 25–35). The author carefully walks you through the science and history of the process. I am going to just summarize that story by saying that the science of analyzing genomes and dealing with gene tree/species tree relationships has advanced tremendously. Computer analysis, fast sequencing, and genomic science have combined to form a new discipline. In essence, the gene tree/species tree incongruities are rarities, and so the dominant statistical trend of the gene trees represents the valid species/tree relationship. The final conclusion was that Tree 1 with the gorilla branching off first, chimp and human branching off last, is the valid species/tree relationship. This Tree 1 type relationship is verified by about 2/3 of the genome, whereas the remaining 1/3 of the genome is of either a Trees 2 or 3 type of relationship.

26.11   OUR COMMON ANCESTOR

So, the genetic evidence has our species closest to the modern chimpanzee, but which ape, chimp, or bonobo was that common ancestor most like? Adrienne Zihlman, professor of anthropology at the University of California at Santa Cruz, concludes after exhaustive study that the common ancestor of chimps and humans most likely looked like a bonobo, a.k.a. pygmy chimp. She concluded that early hominins, like Lucy, resembled the bonobo in body features more so than they resembled the common chimp. Moreover, humans are socially and sexually more like bonobos than like chimps. This hypothesis surprised me at first because the bonobos are known to have split off from chimps long after we did. However, if the common ancestor was more bonobo-like, then it is the common chimp that has changed the most. We have no fossils of the common ancestor to tell us otherwise, so Professor Zihlman's idea may well be correct.

26.12   SUMMARY

The hominid trichotomy refers to the genetic relationship between humans, chimpanzees, and gorillas. The question was which two species of them are most closely related: chimps and gorillas, chimps and humans, or gorillas and humans? The problem was that different experiments yielded different answers, and it seemed impossible to settle the debate. The resolution came from the following logic: The gene tree/species tree incongruities are rarities, and so the dominant statistical trend of the gene trees represents the valid species/tree relationship. The final conclusion was that Tree 1 with the gorilla branching off first, chimp and human branching off last, is the valid species/tree relationship. This Tree 1 type relationship is verified by about 2/3 of the genome, whereas the remaining 1/3 of the genome is of either a Trees 2 or 3 type of relationship.

CHAPTER 27

THE AGE AND ORIGIN OF OUR SPECIES

CONTENTS

27.1 SCOPE

We are the last surviving species of bipedal hominins after a string of numerous species that lived before us. There have been two or more species or sub-species of hominins living concurrently throughout that 5 to 7 million-years period. In fact, we were not alone as little as 30,000–40,000 years ago when Neanderthals shared Eurasia with our ancestors. Now, we are the last man standing, and we are *Homo sapiens*. Is there any way we can determine how long our species has existed? That is the question addressed in this chapter. We shall see how DNA science attempts to answer the question and how reasonable its answer appears to be.

27.2 DNA IS THE KEY TO THE PUZZLE

How long has our species existed? DNA testing has been used to help answer this question. Originally, two particular approaches had been used: The first approach was to find the age of the woman who would be the most recent common female ancestor to all living people on the Earth. This was done via mitochondrial DNA. The second approach was to find the age of the most recent common male ancestor to all living people on the Earth. This was done via the Y-chromosome in males. These individuals have been named "Mitochondrial Eve" and "Y-chromosomal Adam," respectively. In both approaches, scientists examined DNA from living people who represented populations from all over the Earth, examined the mutational changes to the DNA, and calculated how long ago our common ancestor lived. In more modern times, fast-sequencing machines have made it possible to compare the entire genomes of people living around the world. The so-called "1000 genome project" helped provide the necessary data to undertake this endeavor. The newer effort helped us see the problem in a different light.

27.3 MITOCHONDRIAL EVE

27.3.1 MORE ABOUT mtDNA

Within our cells but outside of the nucleus (i.e., the cytoplasm), hundreds of mitochondria exist for the purpose of supplying energy. In fact, this burning of nutrients by combination with oxygen also helps us keep our bodies warm. These mitochondria have a very different DNA than our hereditary DNA. They are only 16,500 nucleotides long compared with over three billion nucleotides for our nuclear DNA. Another big difference is that mitochondrial DNA (a.k.a., mtDNA) is reproduced asexually and is passed to us only from our mothers.

You and your siblings got your mtDNA from your mother. You and your first cousin got it from one of your grandmothers. If each of us were able to trace our ancestry far back in time, more and more people would find they have a common female ancestor. In fact, all human beings on Earth today have a female ancestor common to us all. Obviously, she lived thousands of generations ago. Since we can track her back in time by using mitochondrial DNA mutations, we shall refer to her as Mitochondrial Eve (a.k.a., mtEve).

There are advantages to using mtDNA instead of nuclear DNA for determining our species age: (i) it is more plentiful, (ii) it reproduces asexually, which avoids the complications introduced by gene swapping, and (iii) it mutates many times faster than nuclear DNA, which means it contains 10 times as many mutations that serve as markers. The more markers the better for this kind of problem. Scientists were thus able to examine the mtDNA of randomly selected people all over the world and used the mutational markers to estimate mtEve's age. The variation of mtDNA between different people can be used to estimate the time back to a common ancestor.

27.3.2 THE WILSON, CANN, AND STONEKING FAMOUS STUDY

Rebecca Cann began the study in 1979 by collecting mtDNA from women of different ethnicity. She used the mtDNA in placentas that were collected from U.S. hospitals. This original work encouraged her to believe that the approach could find the most recent common female ancestor. Mark Stoneking joined the effort in 1981. He was a graduate student also working in Allan Wilson's lab for his PhD. Besides the U.S. hospitals, samples were also collected from Australia and New Guinea. Finally, in 1985, Wilson was convinced that the work was ready to publish, and the article was published in 1987 in *Nature*.

The article concluded that Mitochondrial Eve lived in Africa between 100,000 and 200,000 years ago. Later, research tightened the time span to between 140,000 and 200,000 years. Actually, Wilson and team did not use the term Mitochondria Eve in their paper. The term resulted from the popular media. So what does this mean to the age of our species? It means our species is at least 100,000 years old and probably much older. You should understand that mtEve is not the first mother in the sense of biblical Adam and Eve. Obviously, mtEve herself had a mother, grandmother, and so on. Mitochondrial Eve is simply the most recent woman who is a common ancestor to all people on Earth. There were probably many women before her who were also our common female ancestor.

The testing and retesting was done by Alan Wilson, University of California, at Berkeley and his assistants, Mark Stoneking and Rebecca Cann, over a period of years. In 1987, they finally pinned down the place where mtEve had lived. It was Africa and they narrowed her time to 140,000 to 200,000 years ago. Subsequent estimates of mtDNA Eve's age have differed from this range but are still in the same ballpark.

27.4   Y-CHROMOSOMAL ADAM

Mitochondrial DNA was the best approach for its time, but the Y chromosome has advantages over it. Whereas mtDNA has 16,500 nucleotides, the Y chromosome has 58 million nucleotides. It, therefore, has more information contained in it. Y-Chromosomal Adam (a.k.a., yAdam) became a feasible target once the technology made it possible. The quest for yAdam does use our entire nuclear DNA but only a small segment of it. At conception, one particular chromosome determines what our sex will be. If that chromosome happens to be an X, we become a female and if it happens to be a Y, we become a male. Mothers do not have a Y chromosome to give us, so males get our Y chromosome solely from their dads. Our dads got theirs from their dads, and as far back as you want to go, males got their Y chromosome from their dads. This fact allows us to seek out yAdam, the common male ancestor to all living men on Earth. Like the mtDNA had markers, the Y chromosome also has mutational markers, which allow scientists to backtrack to yAdam. Scientists estimate that yAdam lived 200,000–300,000 years ago. As in the case with mtEve, yAdam is not the first man. He is simply the first man who is the most recent common ancestor to all living men on earth. He had a father and grandfather and so on.

It is not necessary that mtEve and yAdam lived at the same time or that they ever knew each other. They are only like their biblical namesakes in the sense of being ancestral to all living people. They were not the first man or woman and had parents, grandparents, and so on, of their own.

27.5   FOSSIL EVIDENCE

Bear in mind though that mtEve and yAdam are not the founding individuals of our species. They are only the most recent woman and man who are a common ancestor to all living people. Our species could be a lot older than their dates indicate. Moreover, new species do not suddenly appear fully formed. Our species gradually evolved from a very similar older species in a gradual manner. *Homo heidelbergensis* is one possible ancestral species to us.

27.5.1   HERTO REMAINS

So, is there any fossil evidence to support the genetic age estimate for our species? Actually, one particular fossil discovery of *Homo sapiens* shows that our species is at least 160,000 years old. Tim White led the excavation that found two adults and one child in the Afar region of eastern Ethiopia. The fossils have been dated at 160,000 years ago. The team also unearthed skull pieces and teeth from seven other hominid individuals, hippo bones with cut marks on them, and over 600 stone tools including hand axes. Refer back to Figure 14.1 to see a photo of the skull.

27.5.2   OMO REMAINS

Richard Leakey led a team that discovered hominin bones at Omo Kibish in Ethiopia. There were two excavations: one in 1967 and the other in 1974. They are called Omo I and Omo II, respectively. Leakey classified them as *Homo sapien* in 2004, and the layers in which the fossils lay was argon-dated at 195,000 years old. This would make them the oldest known *Homo sapien* remains ever found. Since I first wrote that statement, a recent fossil find in western Africa has made the claim obsolete. The earliest fossil find of our species is now pushed back to 300,000 years.

27.6   NUCLEAR DNA TELLS A MORE COMPLETE TALE

27.6.1   THE LIMITATIONS OF THE MITOCHONDRIAL EVE AND Y-CHROMOSOMAL-ADAM STUDIES

The aforementioned mtEve and yAdam studies were leading-edge technology in their time. However, ultra-fast sequencing and computerized analysis techniques allow us to look at entire genomes these days. We can look at changes in hundreds of millions of nucleotides instead of thousands and that gives us more complete and accurate picture of our genetic heritage. Eugene Harris, in his book *Ancestors in Our Genome*, used the following analogy to illustrate the difference between the molecular length of the mtDNA to that of our entire genome: If a DNA base were one-foot-long, the mtDNA would be as tall as Mount Kilimanjaro, but the genome would be triple the distance from the Earth to the Moon.

Harris also believes that the mitochondrial and Y-chromosome studies can be misleading because they eliminate all of those lineages that link across male and female lineages. In other words, genomes better represent the entire genetic history and thus tell the most accurate account of what actually happened.

27.6.2 WHAT DO SNP'S TELL US

The term "SNP," pronounced 'snip,' is shorthand for "single nucleotide polymorphism." In other words, the nucleotide exists in two different forms within the subject population. For our purposes, a SNP is useful for comparing two different genomes. A SNP is a difference between the genomes at one particular nucleotide. There are 3.2 billion nucleotides to compare, so computers are going to be doing the work. To help calibrate your SNP meter, consider the following facts.

- Humans and chimps have about one SNP for every 100 nucleotides.
- Any two humans have about one SNP for every 1000 nucleotides.

Genome comparisons using SNPs can give us a comparison of the diversity between individuals of a specific area and that can tell us something about our origins and migrations. It turns out that diversity is clearly the highest in Africa and diminishes for areas in proportion to their distance from Africa. That means that our species originated in Africa and migrated into the other continents. We are talking about tens of thousands of years and all the many generations that entails. The farther from Africa they traveled, the fewer were their choices of available mating partners, and that is why diversity diminishes with distance from the original location.

27.6.3 MICROSATELLITES TELL A SIMILAR STORY

Recall that DNA sequences are composed of the four letters, A, T, C and G. Within the DNA sequences, microsatellites can be detected. A microsatellite is a string of letters that tends to repeat from twice to multiple times. For example, the microsatellite "AGAT" might be detected as a doublet "AGATAGAT," a triplet "AGATAGATAGAT," and even larger multiples. Microsatellites are a great target for computers to search for within genomes because they mutate frequently with the result that microsatellites tend to get longer with the arrival of new generations.

Scientists have studied the results of microsatellite size and distribution within the genomes from individuals in different regions of the world and confirmed the results of analyzing SNPs. Microsatellites' size distribution in a population group is a measure of diversity. They saw fewer and fewer differences as they traveled farther from Africa. And again, Africa has the greatest diversity and that diversity declines with distance from Africa.

27.7 SUMMARY

The earlier work toward determining the age of the *Homo sapiens* species relied on mitochondrial DNA analysis and Y-chromosome analysis. mtDNA analysis involved females exclusively, and the common female ancestor was dubbed Mitochondrial Eve with an age of between 140,000 to 200,000 years. Y-chromosomal Adam was the common male derived from analysis of Y-chromosomal markers from living males around the globe. His age was estimated to be between 200,000 and 300,000 years. Actual *Homo sapiens* fossils of our early ancestors fit into these time estimates: Tim White found fossils dated at 160,000 years, and Richard Leakey found fossils that date at 195,000 years. A recent fossil find pushes the earliest date back to 300,000 years.

Advancements in scientific techniques allow us to compare entire genomes rather than the short DNA segments of mitochondrial DNA and Y-chromosomes. Examination of entire genomes should yield a more accurate picture of our genetic history. Computer programs can look for

SNPs and microsatellites as they compare genomes. The more of these polymorphisms found or repeated sequences found, the greater the diversity in the test population.

CHAPTER 28

OUT OF AFRICA

CONTENTS

28.1   SCOPE

This chapter is concerned with the migrations that took our *Homo sapiens* ancestors out of Africa and into every region of the globe, save Antarctica. Human bones, stone tools, campfire remains, and other physical evidence help substantiate our claims, but the most valuable information is genetic. Our prehistorical journeys are recorded in our DNA. Modern man has been clever enough to coax the stories from the genetic code. The detailed story of these migrations is a book-length endeavor, and those books do already exist. My goal is in this chapter is to explain how the code-breaking magic was done and to acquaint the reader with some parts of the odyssey.

28.2 WHAT WE ALREADY KNOW

28.2.1 HISTORICAL AND ARCHEOLOGICAL FACTS

Historical documents can only take us back so far because writing was first invented a mere 5000 or so years ago. The *Out-of-Africa* migration precedes the dawn of writing by approximately 45,000 years. Prior to and concurrent with early writing, we rely on archeology to unravel the past. The artifacts and recovered scripts tell us that there was a Copper Age, followed by a Bronze Age and an Iron Age. Stone tools precede the metal ages all the way back to 2.5 million years ago.

Archeologists see differences in the stone tool-working cultures of the different ages to distinguish them. The major breakdown of the Stone Age is the Neolithic (most modern), Mesolithic, and Paleolithic periods, the last one being our oldest known stone tool technology. A series of refinements in stone-working techniques have occurred during the last 50,000 years as an innovative spirit presents itself. Prior to that, there were very long periods where stone toolmaking techniques and products remained unchanged. From *Homo habilis* living 2.5 million years ago to modern *Homo sapiens*, we of the Homo lineage have been Stone Age people for most of our time on this planet.

28.2.2 POPULATION ESTIMATES FOR HOMO SAPIENS

Estimates of our species' population increase over time and give a valuable backdrop to our examination of human migration. It is estimated that 150,000 years ago, our species had about 20,000 individuals all together. One hundred thousand years ago, the earth was in a warm period and our species had penetrated as far north as Israel. The Neanderthal people, also occupying the area, may have prevented our further advances. By 65,000 years ago, our population had grown to, perhaps, one million people. It was about this time when serious migrations out of Africa first began. Twelve thousand years ago, we had about six million people on the planet. At this time, agriculture and herding of animals was gradually beginning to replace nomadic hunting and gathering as a means of sustenance. By 1 A.D., our population skyrocketed to 250 million people in the world. When I was a youngster, I remember talk of the world population hitting 4 billion people. Today, there are about 7 billion people on the planet.

28.3   MULTIREGIONAL THEORY VERSUS OUT-OF-AFRICA THEORY

One of the dominant theories of modern human origins was the multiregional theory, where it was thought that modern humans descended from the archaic humans in different regions of the world. This would have modern Europeans descended from Neanderthals and modern Asians descended from *Homo erectus*. The rationale for the multiregional theory is the marked physical differences between Asians, Negroes, and Caucasians, etc. Proponents argued that it must have taken many thousands of generations for those changes to take place. Today, thanks to information drawn from our DNA, the majority of Western scientists have abandoned the multiregional theory in favor of the Out of Africa theory.

Now it is known that all modern humans are but one species that originated in Africa and that populated the world by migrating out some 50,000 to 100,000 years ago. The Chinese are not all that happy with the "Out of Africa theory. They had made the multiregional theory part of their political thinking. Their line of descent was believed to be from the Peking man, a *Homo erectus* fossil man, who lived there 400,000 years ago. DNA testing has totally undercut their premise, yet many of them do not want to hear about it.

Our species was not the first to leave Africa and settle in different parts of the world. *Homo erectus* beat us to it. His fossils were found in both Peking and Java. A primitive form of *Homo erectus* was discovered in a fossil site at Dmanisi, Georgia, dated at 1.8 million years old. So we were not the first species to migrate out of Africa. Our Out of Africa migrations are much more recent and involve only our species, *Homo sapiens*. We absorbed or replaced all other hominin species in the process of our expansion.

28.4   WHAT DNA TESTING CAN TELL US

28.4.1   GENETIC MARKERS

Those migration times and paths of prehistoric *Homo sapiens* might seem to be lost for all eternity, but don't underestimate human curiosity, persistence, and innovative thinking. Scientists have found an ingenious way to learn what those migration times and paths actually were. They examined the DNA of living people sampled from around the world. Now, we know that harmless mutations occur at a somewhat regular rate over time. An example of a harmless mutation is a change in the code letters of our junk DNA. The change might be a deletion, an addition, or a swap in the code letters. Harmless mutations can also occur within the code for genes as well. When genes exist in two different versions, scientists call that a polymorphism. The two versions of a gene are called alleles, and when alleles differ at a single nucleotide, it is called as single nucleotide polymorphism (SNP).

A mutation leaves a so-called marker on our DNA. We also know that as generations go by, newer markers are left on our DNA. Identification of DNA markers using computers and then establishing the chronological order of the markers makes it possible to estimate an age for them. Applying these techniques to regional populations allows us to establish the migration routes of early humans. We have already seen how Mitochondrial Eve and Y-Chromosomal Adam were found as the most recent common female relative and most recent common male relative for our species. The same processes have been used to identify the most recent common ancestor at various points in the migration. We call these genetically new populations by the term "Haplogroups." Haplogroups have been developed based on mtDNA and Y chromosomal results.

28.4.2   THE SCIENTIFIC PROGRESSION

Professor Luigi Luca Cavalli-Sforza was the pioneer in examining genetic differences in populations. At first, he worked with blood type difference (A, B, and O) and then with protein differences. When it became technically possible to sequence mtDNA, tremendous progress was made. The next big jump in technology was the ability to sequence the Y-chromosome and to identify the polymorphisms it contained. This added greater resolution to the migration studies. The Y chromosome should provide better resolution; it is 50 million nucleotides long compared to the 16,600 nucleotides in mtDNA. Most of the material in this chapter is based upon discoveries using mtDNA and Y chromosomal sequencing. Nowadays, the fast-growing field of bioinformatics might use the entire genome (3.2 billion base pairs) to extract even more information.

Next, we discuss the actual migrations out of Africa. The source that has been most useful to me is *The Journey of Man* by Spencer Wells. If you would like to get a thorough account of the human migration story, I highly recommend this book to you.

28.5   THE ODD CASE OF AUSTRALIA

28.5.1   MODERN HUMANS IN AUSTRALIA

Humans were in Australia at least 45,000 years ago, and there are fossils to prove it. Human artifacts may push the date back to at least 60,000 years ago. In addition, there was an extinction of the mega-fauna in Australia occurring at least 45,000 years ago. The extinction is likely due to man's hunting there. In order to get to Australia from the nearest landmass, our ancestors had to cross about 80 miles of water even with the lowered seas of that period. We may have navigated the oceans in some crude crafts, although no evidence of the craft survived. Our migrations toward Australia were probably along coastlines back then, but the routes cannot be examined today because oceans are much higher, and those ancient shores are under water. New Guinea was accessible by land bridges from Australia until about 8000 years ago. The same people populated both New Guinea and Australia.

28.5.2   HUMAN PRESENCE IN AUSTRALIA IS SURPRISING

Australia is about 7000 miles from northern Africa, yet humans were there some 10,000 to 20,000 years before they showed up in Europe or most of Asia. The genetic evidence says modern humans left Africa about 50,000 years ago, but fossil evidence tells us that Australia was occupied by humans about this same time or even earlier. Adding to the mystery is the fact that bodies of water had to be crossed to make the journey possible. The genetic evidence for the Out of Africa date come from a Y-chromosome marker called M168. The first man with this mutation is the most recent common male ancestor to all non-Africans alive today. We might call him Eurasian Adam. He is dated at between 31,000 and 79,000 years ago. Eurasian Eve (a.k.a. Haplogroup L3) based on mtDNA is dated at between 50,000 and 60,000 years ago.

28.5.3   AUSTRALIAN GENETIC MARKERS

The humans who made this journey are the ancestors to the Aborigines living there today. What do we know genetically about these native Australians? Nearly 100% of the natives living today carry a rare branch mutation from Eurasian Eve or L3. On the male side, M168 begat a branch mutation called M130. The M130 marker is virtually unknown in Europe but comprises a majority of the Australian markers. These M130 males and M females, who made the journey out of Africa to Australia, left offspring with the same markers along a coastal route through Arabia, Persia, India, and South Asia. The M-130 marker is our coastal migration marker.

These migrations occurred during the Ice Age, when oceans were significantly lower. It was possible to reach New Guinea by coastal routes then. Most anthropologists doubt that these hunter-gatherers were capable of building a seagoing boat to cover the 90 kilometers of ocean between New Guinea and Australia at that time. Tsunamis strike this part of the world at times. Perhaps several men and women survived being swept out to sea and made their way to the Australian shore.

28.5.4   A POSSIBLE SOLUTION TO THE PUZZLE

Although the Out of Africa migration according to genetic evidence began about 50,000 years ago, perhaps there was an earlier migration much earlier along the coastal route to New Guinea. There is evidence based on stone tools for a migration 100,000 to 125,000 years ago into the Arabian Peninsula by crossing the Red Sea when it was at a low-water level. These migrations may explain how modern human teeth dated at 126,000 years happened to be in southern China (Figure 28.1).

28.6   FROM AFRICA TO THE MIDDLE EAST TO ASIA

28.6.1   THE GENETIC TRAIL

The information from Y-chromosomes had told us that M168 was the common male ancestor to all non-Africans. Among his descendants, two new

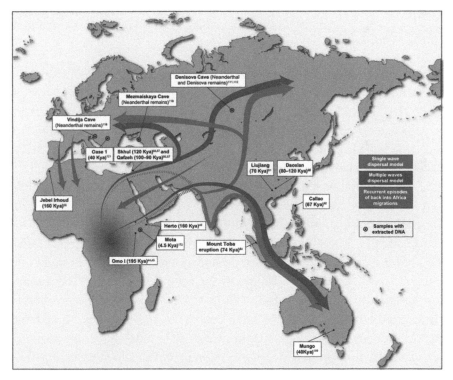

FIGURE 28.1 Putative migration waves Out of Africa (Courtesy of Wikimedia Commons by Saioa López, Lucy van Dorp and Garrett Hellenthal [CC BY 3.0 (http:// creativecommons.org/licenses/by/3.0)], via Wikimedia Commons).

mutations appeared, which have become useful markers to tracking our ancestral migrations. One of them was the coastal marker, M130, which we saw is dominant in Australia. The other new marker is M89. It is on the main line of human migrants headed for Eurasia and has been called the Middle East Marker. The age of M89 is between 30,000 and 50,000 years. The area is the Levant, the land area at the eastern end of the Mediterranean Sea, which encompasses Egypt, Israel, Jordan, Syria, and part of Turkey. It is interesting that M130 markers are not found in Africa, yet M89 markers are found in north Africa.

A new marker on the M89 lineage was designated M9, and its descendants were dubbed the Eurasian clan. This marker occurred in the vicinity of Iran about 40,000 years ago. The M9 descendants are important folks because their descendants ended up in the far corners of the Earth.

28.6.2 MOUNTAINOUS BARRIERS

As the Eurasian clan moved eastward, they were stopped by an impenetrable barrier, a continuous wall of high glacial mountains. The wall consisted of the Himalaya Mountain range, which transverses Pakistan and India in a southeasterly direction and the Hindu Kush mountains of Afghanistan and the Tien Shan mountains of Kyrgyzstan running in a northeasterly direction. The choice was to travel south into India or north into central Asia. Those that traveled south acquired a marker designated M20, which is common in southern India today. Half of the males have it. The M20 so-called Indian clan lived about 30,000 years ago.

Meanwhile 35,000 years ago, the northbound M9 migrants also acquired a new mutation designated M45 in central Asia. The mutation is unique to this area, so we call them the central Asian clan. The migration eastward continued probably via a route through southern Siberia. These people made it all the way to east Asia, and they have a distinct marker designed M175. The people in east Asia today have this marker. For example, it is present in 30% of Korean men. The story is not this simple though. It turns out that the coastal marker M130 is present in east Asians too. It is found in half the Mongolian males. As would be expected from their route described earlier, M130 is older in southern Asians than in northern Asians.

28.7   FROM CENTRAL ASIA TO EUROPE

28.7.1   MODERN HUMANS IN EUROPE

The first time that our ancestors saw Europe was about 45,000 years ago. We had to enter an area already occupied by our sister species, the Neanderthals. By 15,000 to 20,000 years later, the Neanderthals were all dead. It seems that we were better at problem-solving than they were and that eventually made the difference. For example, we invented the sewing needle and made superior garments to protect ourselves from the Ice Age cold periods. We had superior weapons and tools for hunting too. We may have depleted their traditional food source.

Our early ancestors in Europe have a special name. They are called the Cro-Magnons, after a cave in France. About the time that the Neanderthals were dying out, our ancestors left beautiful colored paintings in the caves of France. Portraits of animals, now extinct, were a common theme. A much later wave of migration into Europe happened after agriculture was developed in the Middle East (Iraq). Farmers, looking for fresh farmland, gradually spread throughout Europe.

28.7.2   HOW DID WE GET TO EUROPE?

The genetic evidence says that we did not simply head west from the Middle East. There is a low frequency of the M89 marker in Europeans. The marker found most commonly in European men is M173. It is at very high frequency (90%) in Spain and Ireland. Through the examination of microsatellites, it was learned that the M173 mutation is about 30,000 years old. Being the oldest marker in Europeans, it defines the European clan. M173 is a branch off the M45 lineage, which had originated in central Asia. So, it appears that these steppe hunters had followed reindeer herds in a westerly direction over generations, and it had led them into Europe.

28.8   FROM ASIA TO THE AMERICAS

28.8.1   MODERN HUMANS IN THE AMERICAS

When did humans first set foot in North or South America? That question is still debated. We think that it was fairly recent, say less than 20,000 years ago. We are a little more definite about how the first humans got here and where they came from. We think that Northeast Asians crossed the Bering Straits during the last Ice Age. In other words, humans living in Siberia found a route into Alaska. Perhaps they were hunting big game and were led across as they chased a herd of mammoths.

Remember that this migration happened during the Ice Age, and Siberia was very cold at this time. Even today, the night temperatures can fall to −70°C. These humans had adapted to a brutally cold climate, where plants were very rare and subsistence depended on hunting and fishing. Archeological evidence tells us that although southern Siberia was occupied as early as 40,000 years ago, northern Siberia was not occupied until 20,000 years ago. It took a long time to adapt to such a severe lifestyle.

28.8.2   GENETIC EVIDENCE

The marker found in Native American males is M3. It is at 90% frequency for South Americans and at 50% for North Americans. This would seem to jive with mtDNA results on Native Americans showing there were two migration waves, one populating both American continents and the second only populating North America. The problem is that M3 is not found in Asia. Apparently, this marker developed in North America at an early point in the first migration. Eventually an intermediate marker M242 was found that ties M45 and M3 together. M242 is common in Siberia and was dubbed the Siberian marker. Thus, we have genetic proof that Native Americans originated from Siberian ancestors. While these tundra dwellers were able to move back and forth between the continents as early as 20,000 years ago, they were prevented from traveling south on the American continent due to an extensive ice field. By 15,000 years ago, there would have been enough melting to allow access to the rest of North America and then South America. Like in Australia, the larger animals went extinct soon after the humans arrived. The mammoth, mastodon, and giant ground sloth went extinct about 10,000 years ago.

28.9   FROM ASIA TO THE PACIFIC ISLANDS

28.9.1   MODERN HUMANS IN THE PACIFIC ISLANDS

When I was a boy, I saw my Norwegian-American parents get all excited about the adventures of Norwegian Thor Heyerdahl. It turns out that Thor believed that people from South America, not Asia, had settled the South Pacific islands. After all, the winds and currents traveled from east to west. To prove that South American migration could have been done, Thor built an ocean sailing vessel. It was a balsa wood raft, made without using any modern materials. He called it the Kon-Tiki. Moreover, he did sail successfully 4000 miles to the Tuomotu Islands near Tahiti. His remarkable voyage was the talk of the times. Unfortunately for Thor, DNA testing can undo some of the most staunchly held theories. We now know that the island settlers were remarkable sailors from South Asia and Melanesia. The M130 marker is common among Island males. They used their outrigger canoes and an ability to navigate by the stars to settle on these islands. They brought their families, cattle, and seeds along with them. They settled Fiji 3000 years ago (about 1000 B.C.), Easter Island about 300 A.D., and New Zealand about 800 A.D. Nowadays, newcomers are changing the genes of these islanders through interbreeding. Hawaii is the ultimate example of a mixed gene population. The original Hawaiian DNA is now mixed with Chinese, Japanese, English, Philippine, and many other DNAs.

28.10   SUMMARY

Two theories competed prior to the advent of DNA science in paleoanthropology: The multiregional theory and the Out of Africa theory. The multiregional theory, which proposed that modern humans descended from more primitive species in each region, made sense in view of the physical differences between the major races. The Out of Africa theory proposed that modern humans migrating out of Africa displaced the primitive human species of the world. Application of DNA science and technology to tracking the migrational paths out of Africa led credibility to that theory.

The Out of Africa migration is thought to have begun around 50,000 years ago. However, the first population of Australia seems to have occurred ahead of schedule. There is evidence based on stone tools for a migration 100 to 125 thousand years ago into the Arabian Peninsula by crossing the Red Sea when it was at a low-water level. Future finds may further substantiate this earlier migration.

Scientists first used blood types, then mtDNA, and finally Y-chromosomes to track the Out of Africa migrations. Excellent books have been written on the subject. By testing native peoples from around the globe, genetic markers have been identified, which are first encountered at points along the journey. These markers were dated, and now we have a good idea of the chronology of this so important migration.

Table 28.1 summarizes the relationships between Y chromosomal haplogroups identified in this chapter. Each group has a characteristic mutational marker, which originated at the location indicated and at the time indicated.

TABLE 28.1 Y Chromosomal Haplogroups in "Out of Africa" Migrations

Haplogroup	Name	Comments	Previous	Next
M168	Eurasian Adam	31 K to 79 K years ago in Northeast Africa		M130, M89
M130	Coastal Route Clan	Australia, Mongolia, Siberia, America, and Pacific Islanders	M168	
M89	Middle East Marker	30 K to 50 K years ago, The Levant	M168	M9
M9	Eurasian Clan	40 K years ago; Widely dispersed group	M89	M20, M45
M20	Indian Clan	30 K years ago, India	M9	
M45	Central Asian Clan	35 K years ago	M9	M175, M173
M175	East Asian Clan	Korea, etc.	M45	
M173	European Clan	30 K years ago		
M45	70% of S. Englishmen			
M242	Siberian Clan	20 K years ago	M45	M3
M3	American Clan	90% of S. Americans		
M242	50% of N. Americans			

CHAPTER 29

NEANDERTHAL-HUMAN INTERBREEDING

CONTENTS

29.1   SCOPE

Once it was known that both Neanderthals and modern humans must have encountered one another in Europe and the Middle East some 40,000 years ago, many anthropologists have speculated that the two species might have interbred and had children who were part Neanderthal and part modern human. Whether it ever happened or not seemed like a question that could never be answered. However, the discovery of DNA sequencing offered new hope. To get to the point where the human-Neanderthal interbreeding problem could be tackled using their DNA, it was first necessary to develop the technology of testing ancient bones. Neanderthal bones were many tens of thousands of years old. DNA tends to decompose over time so there was a question of whether it was even possible. The technology for successfully sequencing such old DNA did not exist and needed to be invented. Fortunately, there was a pioneering scientist with the goal of someday comparing Neanderthal and *Homo sapien* DNA, namely Svante Pääbo (Figure 29.1). This chapter is as much about his story as about the question of human-Neanderthal interbreeding. In fact, his excellent book *Neanderthal Man. – In Search of Lost Genomes* is the dominant source for the material in this chapter.

29.2   THE SETTING AND THE QUESTION

For thousands of years, two different types of humans lived as hunter-gatherers in the same geographical area. One of these groups was the Neanderthals, a group that had survived in Ice Age Eurasia from over 120,000 years ago until they went extinct about 30,000–40,000 years ago. The other group was the Cro-Magnon, our ancestors. They came out of Africa around 50,000 years ago and first appeared in Europe about 45,000 years ago. Scientists and others have long speculated on whether interbreeding between Neanderthals and our ancestors occurred during their time together and, if so, what importance it might have had.

When we examine fossil remains, it isn't hard to tell which were Neanderthals and which were Cro-Magnon. Neanderthals were recognizably different from us in that they had a lower forehead, a prominent bony eyebrow ridge, a much larger nose, and lacked a protruding chin. Their bones were thicker and squatter too. Cro-Magnon, on the other hand, looked very much like modern humans in features, stature, and cultural artefacts. The earliest Cro-Magnons were probably dark-skinned with curly hair since they had been adapted to the sunnier climate of Africa. Neanderthals are thought to have had pale skin, light-colored hair, and blue eyes.

Have we ever found fossils of mixed Neanderthal and Cro-Magnon parentage? Some paleoanthropologists believed that certain fossils were suggestive of Neanderthal-human admixture due to the mixed features of those fossils, but others disagreed, and it seemed that the question might never be answered. Now it appears that DNA technology might have resolved the question. It has become a more and more powerful tool for probing into the unrecorded past. The task of resolving the admixture question is the subject of this chapter.

29.3   SVANTE PÄÄBO, PIONEER OF RESURRECTING ANCIENT DNA

The amazingly rapid development of DNA science is due to the dedication and brilliance of many outstanding men and women. One unique area of DNA science is the area of identifying the DNA of long dead animals, including the more recent hominins. This includes animals that have been extinct for some time. One of the difficulties of trying to identify ancient DNA is that decomposition occurs and that other microorganisms may have invaded the targeted DNA, thereby contaminating it. That is why using painstaking precautions to avoid contamination of the ancient DNA with a different DNA are very important. One common contaminant is modern human DNA from people handling the samples.

FIGURE 29.1   Svante Paabo in 2014 (Courtesy of Wikimedia Commons by Jonathunder (own work) [GFDL 1.2 (http://www.gnu.org/licenses/old-licenses/fdl-1.2.html)], via Wikimedia Commons).

Perhaps the most prominent pioneer in resurrecting ancient DNA is Svante Pääbo and his teammates. One of his great early accomplishments was to identify the mitochondrial DNA of Neanderthal man. Pääbo still greater accomplishment was identifying the nuclear DNA of Neanderthal man. One reason why nuclear DNA is harder to analyze is that while mitochondrial DNA is only 16,500 nucleotide pairs long and plentiful, nuclear DNA is 3.2 billion nucleotide pairs long and difficult to extract. The numerous obstacles he encountered and ingeniously overcame make a fascinating story.

29.4   A BREAKTHROUGH PROCEDURE FOR ANALYZING ANCIENT DNA

Contamination was one of the most serious problems in extracting and sequencing ancient DNA. In the early 1990s, Pääbo team evaluated a procedure developed by Erika Hagelberg and J.B. Clegg. The essence of the procedure is that DNA binds to silica powder under conditions of high salt concentrations. The bound DNA can then be washed to remove contamination. Finally, the cleaned DNA can be released from the silica by lowering the salt concentrations. This procedure made it possible for Pääbo team to extract, purify, duplicate, and analyze the mtDNA from four Siberian mammoths, which were between 9700 and 50,000 years old. During this period, much of the laboratory work was done by Matthias Hoess. In 1996, they also sequenced the mtDNA of an extinct Patagonian giant ground sloth, *Mylodon*, and showed by comparison with living sloths that it was more closely related to the two-toed sloths than to the three-toed sloths. Further investigation showed that *Mylodon* and other sloths belong to a group of animals that include armadillos and anteaters. This group of mammals started to diversify over 65 million years ago when the dinosaurs still walked the Earth.

29.5   THE ICE MAN

In 1993, Pääbo was asked to analyze the DNA of the Ice Man, the 5300-years-old mummified body, which had been found in the Alps in

1991. The lead person working on this assignment was Oliva Handt, who had experience working DNA analysis of long dead humans already. She had been working with Native American remains. The tissue samples presented a very challenging problem because they were highly contaminated with mtDNA from humans, who had handled the Ice Man corpse. The team employed elaborate procedures, which involved amplification of single DNA fragments. Knowing that DNA gets broken into smaller and smaller fragments over time, they isolated the most fragmented and most abundant mtDNA belonging to one individual and deemed these fragments to be the true Ice Man mtDNA. These fragments were matched for overlapping segments, and a 300-nucleotide sequence was obtained. The Ice Man mtDNA only differed from modern European man in two positions. This result was expected, considering the short 5300-years difference.

29.6   NEANDERTHAL MITOCHONDRIAL DNA

During the sequencing of Neanderthal mtDNA, Pääbo team employed a process known as polymerase chain reaction (PCR) to vastly increase the sample available for testing. The process amounts to capturing the DNA fragment between two primers and making a duplicate of it. Repeating the process through forty cycles can result in a trillion copies of the DNA fragment. It was also standard practice to clone the fragment in bacteria, have it reproduced to a million copies, and then sequence the cloned copies in order to test for contamination. Extraordinary measures were employed to minimize contamination including: clean rooms, sterile gowns, gloves, face shields, etc. In 1996, by using these elaborate procedures, Pääbo team was able to sequence a segment of the Neanderthal mtDNA.

29.7   THE FIRST NEANDERTHAL mtDNA EVER IDENTIFIED

Special techniques were used to match up overlapping sections of the fragments. Eventually, the team had produced a 379-nucleotide sequence of the Neanderthal mtDNA that was of special interest because it was the most variable part of the mtDNA. Now the Neanderthal mtDNA could be compared to modern human mtDNA. When the mtDNA from 2,051 present-day people were compared to each other, seven positions varied between any of them on average. The Neanderthal mtDNA showed 28 positions varying. The 510 Europeans in the group might have been expected to be most similar to the Neanderthals, but the average difference was still 28. Asians and Africans also were different by 28 positions. These data tend to confirm that the sample of mtDNA from the Neanderthal fossil was Neanderthal mtDNA and not human mtDNA, which has been an unwanted contaminant. Later, Matthias Krings sequenced another section of the Neanderthal mtDNA, and when compared with the matching section in modern humans, it indicated that our common ancestor lived about 500,000 years ago. The Neanderthal mtDNA comparison with human mtDNA also suggested that little to no interbreeding occurred between Neanderthals and our direct ancestors.

29.8   VARIANCE IN mtDNA BETWEEN INDIVIDUALS

Meanwhile, Pääbo had arranged for samples to be received from other Neanderthal individuals so that all the findings didn't hinge on a single individual. They selected a fossil dated at 42,000 years old and sequenced its MtDNA. It was a close match to the first individual. Independent researchers also confirmed the mtDNA sequence.

Pääbo was interested in determining the variance of the mtDNA segment between individuals and how it compares humans and great apes. They had the mtDNA from three Neanderthal individuals, so they used three individuals as a basis of comparison. The results are in Table 29.1.

TABLE 29.1   Variance in mtDNA Between Individuals

Sampled	Variance
Neanderthal Man	3.7%
Modern Humans	3.4%
Chimpanzees	14.8%
Gorillas	18.6%

Such low variance is indicative of small populations. Although the sample size was very small, it was tentatively concluded that Neanderthals, like us humans, had expanded from a small population.

29.9 EARLY EXPERIMENTS IN SEQUENCING ANCIENT NUCLEAR DNA

29.9.1 CAVE BEAR BONES

Pääbo added Alex Greenwood, a new post-doc from the United States, to the team. His assignment was to see if nuclear DNA could be identified from ancient animals. This activity was in preparation of attempting to do it with Neanderthal bones. They began with 30,000-years-old cave bear bones taken from the same caves that contained Neanderthal fossils. The cave bear bones were more plentiful and less precious. However, their numerous attempts to find and amplify nuclear cave bear DNA were not fruitful. The main reason for failure is that compared to the plentiful mtDNA genomes, nuclear genomes are a rarer and more elusive target.

29.9.2 A MAMMOTH'S FROZEN TOOTH

They switched to an easier fossil, a frozen mammoth tooth. If it were possible to find nuclear DNA in an extinct species, this would be the easiest one because DNA decay had been prevented by the freezing temperatures of its environment. Instead of trying to find nuclear DNA itself, they sought to find the ribosomal equivalent of it. Specifically, they sought a part of a gene that was abundant in the cell attached to ribosomes. The gene is called 28S rDNA. Greenword was successful in amplifying the ribosomal gene and sequencing the genetic code it held. Ah! Sweet success! A 14,000 years old mammoth tooth had surrendered its DNA code, and history had been made again by Paabo's team. These were the first nuclear DNA sequences ever determined from a long extinct mammal.

Greenwood then tried to get ribosomal DNA from a very promising cave bear bone. He was successful but just barely. There appeared to be a huge difference between nuclear DNA preservation in permafrost versus in limestone caves.

29.9.3   CAVE BEAR BONES REATTEMPTED

Great scientists are great because they are persistent and tenacious. Pääbo with the assistance of Hendrik Poinar, had a breakthrough that made it possible for him to resurrect ancient nuclear DNA, such as cave bear bones. The exact story of how this occurred is highly technical, so I will quickly summarize it. However, if your organic chemistry is good, then you might want to read the detailed story in Pääbo book *Neanderthal Man* (pp. 105–107).

The essence of the story is that with time, some of the DNA has chemically reacted to become something else (i.e., per the Maillard reaction). The unknown substance is not recognized as DNA by the procedures that Pääbo had been using. A clue to its presence is that the substance gives off a blue fluorescence in UV light as do Maillard reaction products. Now, it wondrously turns out that a reagent exists that can reverse the degradation process and resurrect the DNA again. The chemical reagent was given the designation "PTB." When Pääbo and Poinar added PTB to extracts from ancient samples of cave bears and Neanderthals, they saw improvements in detection. With perseverance and ingenuity, they had turned discouragement into hope again. Discovering the identification of nuclear Neanderthal DNA was viable again.

29.10   THE SECRETS OF NEANDERTHAL NUCLEAR DNA UNLOCKED

I am going to jump ahead to the final stage of this challenging quest. For the first time in the history of the world, the nuclear DNA of Neanderthal-type humans was revealed for mankind to behold. The roadblocks encountered in this journey were monumental and would have discouraged any of us. For one, their source of Neanderthal bones containing useful DNA disappeared due to a political change. No matter how discouraging the circumstance, Pääbo was not willing to be stopped. He used every tool at his disposal to overcome adversity, be it using his academic knowledge, conducting endless experiments, using his charm, using his personal connections, and betting that solutions would miraculously appear. When I finished this section of his book, I acquired a deep admiration for this multidimensional man. He is the definition of inspirational.

Constructing the Neanderthal genome was a process of sequencing the small available fragments and working out a method of properly stringing the fragments together. Human DNA was one reference to use as a guide, but one had to identify differences between Neanderthal and human DNA because that was the goal of the effort. Chimpanzee DNA was another reference point equidistant in time from both Neanderthal and human DNA. A third reference was an imaginary genome designed to contain the DNA of the human-chimp common ancestor. The mathematical process needed to resolve these mapping puzzles was huge. To the rescue was the loan of 256 powerful computers by the Max Planck Institute.

As the Neanderthal genome began to materialize, Pääbo realized that he needed to add the right kind of brainpower to the team; in particular, he needed population genetists to help determine if Neanderthals were closer genetically to humans of particular locations, like Europe, for example. He was able to add very strong talent to the analytical problems.

29.11   WHEN DID WE SPLIT FROM NEANDERTHALS?

Pääbo team performed a genomic analysis that was designed to establish when humans and Neanderthals split into distinct groups. First, they determined all the positions in human DNA where we had a derived base. In other words, the base at that position is different than what the same position is for a chimpanzee. Since the chimpanzee is ancestral to us, the change has occurred during our separate evolution. Next, they looked at these two million plus sites in the Neanderthal genome and determined what percentage of these sites were the same as human DNA. It turned out to be 18%.

They calculated the human-Neanderthal split time assuming a chimpanzee divergence of 8.3 to 5.6 million years ago. In Table 29.2 shows the estimates of the human-Neanderthal split based on the 2012 Neanderthal genome and also on the far more accurate 2014 Neanderthal genome.

TABLE 29.2   Time Estimates of Human-Neanderthal Split

Nuclear genome	High Estimate	Low estimate
2012 Neanderthal	440,000 years ago	270,000 years ago
2014 Neanderthal	383,0000 years ago	275,000 years ago

Note: Dates from Wikipedia - Upper Paleolithic, https://en.wikipedia.org/wiki/Upper_Paleolithic.

29.12   DID HUMANS AND NEANDERTHALS INTERBREED?

29.12.1   DEVELOPING A HUMAN GENOME DATABASE FOR COMPARISON

Although the earlier study of Neanderthal mtDNA gave no indication that interbreeding had occurred, the nuclear DNA evidence was not so clear. As more and more of the Neanderthal genome filled in, the probability of interbreeding increased. Part of the analysis was dependent upon comparison with human genomes that had been sequenced by outsiders to Pääbo team. To eliminate doubt, they decided to generate their own human genome data using candidates from France, Papua New Guinea, West Africa, South Africa, and China. Testing began in 2009.

29.12.2   WHAT WAS LEARNED FROM THE GENOME COMPARISONS

*   **Result #1** — One result of the DNA testing and data analysis was that the point of divergence between Neanderthal and human species was not that great when compared with the wide variation amongst living humans. This realization came after learning that the San (South Africa) candidate's DNA indicated a point of divergence with the reference human genome of 700,000 years. The other five

human candidates diverged at about 500,000 years. We know that all modern humans are one species and can breed with each other, so it seems more likely that Neanderthals and humans could have produced offspring too.

- **Result #2** — The next result was that Neanderthals were genetically closer to non-Africans than to Africans. This is logical because Neanderthals were never in Africa.
- **Result #3** — Finally, there seemed to be a small but discernible genetic contribution of Neanderthal DNA in non-African genomes. In other words, after leaving Africa, humans interbred with Neanderthals, and the evidence is in their DNA.

29.13   MORE ON HUMAN-NEANDERTHAL INTERBREEDING

Pääbo team also looked into gene flow resulting from interbreeding. They learned that the gene flow is predominately one way: from Neanderthal into modern humans. This implies that the Neanderthals were socially dominant and that Neanderthal men were impregnating human women.

Monty Slatkin employed modeling of human and Neanderthal population histories to determine the percentage of Neanderthal DNA in living humans. His calculations arrived at a value of between 1 and 4% for people of European or Asian ancestry. A different calculation done by Pääbo's team arrived at a value of between 1.3 and 2.7%. So it is safe to say that non-Africans have something less than 5% Neanderthal DNA in their genomes.

29.13.1   WHERE AND WHEN INTERBREEDING OCCURRED

Modern human bones uncovered at Skhul and Qafzeh caves in Israel are over 100,000 years old. Neanderthal skeletons dated to 45,000 years old have been found in nearby Tabun Cave and Kebara Cave. So it is likely that humans and Neanderthals encountered one another over the tens of thousands of years that they both dwelled in this area of the Middle East. Pääbo thought this area and time period would have been the most likely place and time for interbreeding to occur. Another point of interest is that the stone tool culture used by both was identical. This suggests that they interacted with each other.

The situation changed after 50,000 years ago. A new stone tool culture (i.e., Aurignacian) was adopted by our migrating ancestors, and their population was growing rapidly. Projectile weapons were developed during this period, and as the Aurignacions moved into an area, Neanderthals disappeared from that area soon after. Pääbo suggests that those Aurignacian people carried Neanderthal genes in their genomes and spread their modified genomes to their offspring and others that they encountered during their migrations.

Further analysis suggests that interbreeding must have happened during contacts 50,000 to 100,000 years ago in the Middle East.

29.14 IMPLICATIONS OF INTERBREEDING TO THE OUT OF AFRICA THEORY

The original Out of Africa theory envisioned modern humans migrating out of Africa and displacing more primitive hominins wherever they were encountered. It did not envision assimilation of the more primitive hominins as a result of interbreeding. However, it is now clear that such a process happened with the Neanderthals. Moreover, the modified genomes of our migrating ancestors due to a Neanderthal component would have been carried and mixed into new populations. The possibility that has happened with other archaic humans is also likely. The new Out of Africa theory has integrated assimilation now into its doctrine.

29.15 SUMMARY

Svante Paabo pioneered the DNA restoration and analysis of ancient DNA. In 1993, he was selected to sequence the mtDNA of the Ice Man, a 5300-years-old man who had been preserved by Alpine snow. In 1996, his team sequenced the mitochondrial DNA of Neanderthal man. This is one of his historic accomplishments. Comparison with human mtDNA placed our common ancestor at 500,000 years ago.

Next, he took on the problems of extracting nuclear DNA from Neanderthal bones and sequencing. Trial runs on cave bear bones at first were discouraging. So, he tried an easier task. He tried to sequence the nuclear DNA from a 14,000-years-old mammoth's tooth. But instead of trying to find nuclear DNA itself, he sought to find the ribosomal equivalent of it. Specifically, they sought a part of a gene that was abundant in the cell attached to ribosomes. He was successful and made history as the first person to sequence the nuclear DNA of a prehistoric animal. He then found a reactant that reversed the degradation of DNA and applied the technique successfully to cave bear DNA and Neanderthal bones. Other obstacles were overcome, and the Neanderthal nuclear DNA project was underway. Pääbo added additional experts to the team as this was a challenging undertaking. After a Herculean effort, Pääbo accomplished his greatest undertaking and sequenced the nuclear DNA of extinct Neanderthal bones.

One thing they determined from the Neanderthal genome was the time of a common ancestor with modern man, it was as early as 270,000 years ago and as late as 440,000 years ago. They also learned that Neanderthal DNA is part of the human genome. Up to 5% of our DNA is Neanderthal. The interbreeding probably occurred between 50,000 and 100,000 years ago in the Middle East where the two species sometimes shared territory. The discovery of this interbreeding forced a reevaluation of the Out of Africa theory. Breeding between species or sub-species is a fact that must be included in the theory.

CHAPTER 30

DENISOVAN-HUMAN INTERBREEDING

CONTENTS

30.1 SCOPE

Svante Pääbo the hero of the last chapter, reappears in this chapter. Instead of being like his previous success story resulting from following a defined goal tenaciously, we have instead a case of serendipity. A fantastic opportunity fell into his lap, and he was uniquely prepared to exploit the opportunity. A finger bone from a cave in Siberia was rich in mtDNA, and Pääbo lab was able to sequence it. The mtDNA was unlike Neanderthal or modern man; it was something else. A new species was defined by a small finger bone. The story follows.

30.2 WHAT IS A DENISOVAN?

As far as fossil evidence goes, we know very little about Denisovans. In fact, we probably would never even have known of their existence if we hadn't discovered their ancient DNA. In 2007, Svante Pääbo was wrapping up his effort to decode the nuclear DNA of Neanderthal bones. He had been collaborating with Russian archeologist Anatoly Derevianko on small bones in a Siberian cave, which proved to be Neanderthal, according to mitochondrial DNA analysis. In 2009, Anatoly submitted a miniscule finger bone fragment found in another Siberian cave, namely, Denisova Cave. The mitochondrial DNA in the bone was so plentiful that Pääbo team was able to sequence the entire mitochondrial genome. The cold temperatures of the region may have kept the mtDNA in excellent condition. The test results were startling! It did not match either modern human mtDNA or Neanderthal mtDNA. It was something different from either of them. Where Neanderthal mtDNA varied from human mtDNA at 202 nucleotide locations, Denisovan mtDNA varied from ours at 385 locations. The differences in the mtDNA suggested that Denisovan man and humans had a common ancestor twice as far back in time as when we had a common ancestor with Neanderthals. In other words, our common ancestor with Denisovan man had lived about one million years ago.

30.3   DENISOVAN MOLAR FOUND

In addition to the finger bone, a molar was also found from Denisovan man. A complete mitochondrial DNA genome was sequenced that only differed from the finger bone scan in two positions. Thus, the tooth came from a different individual than did the finger bone. The tooth itself was interesting. It was twice the size of a modern human molar, but it was dissimilar from a Neanderthal molar too. There were differences in the shape of the crown, but the biggest difference was in its diverging roots. The roots of Neanderthal molars are closely spaced and sometimes even fused. The age of the Denisovan fossils is unknown, but there is evidence that the cave was occupied about 30,000 years ago and about 50,000 years ago by Neanderthals and/or Denisovans. The fossils are in this age range. Pääbo favored the 50,000-years-old age for Denisovans.

30.4 DENISOVAN NUCLEAR DNA EXAMINED

Pääbo decided that the next step should be to attempt the sequencing of Denisovan nuclear DNA. He was lucky in this pursuit, because the nuclear DNA was plentiful and in a good state of preservation. Moreover, all of the lessons learned from the Neanderthal genome project could be directly applied to the Denisovan genome project. The first thing he learned from comparing the nuclear DNA was that Denisovans are genetically much closer to Neanderthals than they are to modern humans. In fact, the degree of difference between Neanderthals and Denisovans is like the degree of difference between the most dissimilar modern humans. This conclusion differed radically from the mitochondrial DNA conclusion. Pääbo speculated that a recent cross-breeding might account for the mtDNA results, because one inherits mtDNA from the mother alone. One early result of the nuclear DNA analysis was that no Y chromosome fragments had been found. Apparently, the bone came from a female. It was also found that this Denisovan girl shared more derived SNP alleles with the Neanderthal genome than with modern humans. In other words, Denisovans are more closely related to Neanderthals than to us.

Nuclear DNA analysis puts the Denisovan and human split at about 400,000 years ago. Neanderthals and Denisovans gradually drifted apart, with the final split at about 250,000 years ago.

30.5 DENISOVANS AND CERTAIN HUMANS INTERBRED

The Denisovan nuclear DNA was compared with the DNA of modern humans from five locations, namely: two from Africa, one from Europe, one from China, and one from Papua New Guinea. Denisovan nuclear DNA was most similar to the Papuan DNA. This finding seemed strange because Papua and Siberia are a good distance apart. The analysis was checked and rechecked, and they still came to the same conclusion, so it was decided to add seven additional human genomes to the comparison and see what light that might shed on the mystery. They were: an African Mbuti, a Sardinian, a Mongolian, a Cambodian, a Karitiana (South

America), two people from Melanesia, a second Papuan, and a person from the Island of Bougainville. This testing did not clarify the distant locations issue but did confirm the previous finding that Denisovans and Papuans shared unique features in their DNA. The team also compared Denisovan DNA with DNA from the Human Diversity Panel, which contains DNA data from 938 individuals from fifty-two countries. All seventeen individuals from Papua New Guinea and all 10 individuals from Bougainville stood out as being closer to the Denisovan genome. It now seemed indisputable that somehow Papuans, or more likely their migrating ancestors, had interbred with Denisovans.

A more detailed genetic survey showed that Denisovans had also interbred with Melanesians, Polynesians, Australians, and Philippines. This data seems to bolster the hypothesis that Denisovans encountered Out of Africa humans along a coastal migration route. Pääbo team published their work on the nuclear genome of Denisovan Man in *Nature* in 2010.

30.6   POSTSCRIPT

It was discovered that a tenfold greater yield of nuclear DNA is possible if the double-stranded helix of DNA is made to separate into two strands. The process ultimately increased the recovery of ancient DNA by 10 times. In 2012, Pääbo team reanalyzed the Denisovan DNA but with much greater accuracy. Anatoly Derevianko and his team re-excavated the Denisovan cave and found another huge molar identified as Denisovan.

30.7   SUMMARY

The mtDNA of a Siberian fossil finger joint did not match modern man or Neanderthal mtDNA. Pääbo sequenced the nuclear DNA and learned much more. Unlike the mtDNA results, the nuclear DNA placed the Denisovan fossil quite close genetically to Neanderthals. Nuclear DNA analysis puts the Denisovan and human split at about 400,000 years ago. Neanderthals and Denisovans gradually drifted apart with the final split at about 250,000 years ago. When the Denisovan nuclear DNA was compared with the DNA of modern humans from five locations, namely, two from Africa, one from Europe, one from China, and one from Papua New Guinea. Denisovan nuclear DNA was most similar to the Papuan DNA. Further testing using additional human samples helped offer an explanation. A more detailed genetic survey showed that Denisovans had also interbred with Melanesians, Polynesians, Australians, and Philippines. This data seems to bolster the hypothesis that Denisovans encountered Out of Africa humans along a coastal migration route.

PROBLEM SET FOR PART V

QUESTIONS

Q1. What is mitochondrial DNA and why do we say it is inherited from our mothers?

Q2. Describe "bioinformatics."

Q3. Describe the "hominid trichotomy."

Q4. Three different species/trees were proposed based upon the data in Table 26.1. What are they?

Q5. If someone told you that mitochondrial Eve is the same Eve as mentioned in Genesis of the Bible, what would you say in response?

Q6. SNPs and microsatellites are useful for determining diversity in a given population. Explain.

Q7. Both mtDNA and the Y-chromosome have been used to track the migrations of our ancestors out of Africa. Which of them would be expected to provide better resolution to deciding migration details and why?

Q8. The Himalayas and the Hindu Kush mountains presented an impenetrable barrier to our migrating ancestors. Their possible choices were to travel either northeast or southeast. Where did each group end up?

Q9. The marker found in Native American males is M3. It is at 90% frequency for South Americans and at 50% for North Americans. Why do you think there is a difference?

Q10. Svante Pääbo made a significant historical first using a 14,000-years-old mammoth tooth. What important first had he accomplished?

Q11. Where did Pääbo think Neanderthal-human mating took place and why did he think it?

Q12. Denisovans are genetically closer to Neanderthals than they are to modern humans. How do we know this?

Q13. Denisovans are genetically closer to people from Papuan New Guinea than any other human group. Siberia and New Guinea are a long way apart. What is the explanation?

PART VI

UNIQUELY HUMAN EVOLUTION

In the previous chapters, we discussed the traits that make us different from our chimpanzee cousins. In Part VI of the book, we employ the tools of fossil analysis, evolutionary theory, and DNA science to readdress these traits.

In Chapter 31, we examine the acquisition of bipedal walking by some tree-dwelling apes. The explanations for adopting bipedal walking of three different scientists are presented and compared.

In Chapter 32, we examine the loss of fur by humans. The explanations of two differing scientists are presented and compared.

In Chapter 33, we examine the astounding growth of the brain. Three explanations are presented. It is possible that all three contributed to some degree. However, sexual selection may have been the main driver.

In Chapter 34, we examine the acquisition of speaking and listening and its transition from basic functions into sophisticated languages. Brain experiments have identified regions of the brain with speech defects. The center for the original European language has been found and its mode of spreading identified.

In Chapter 35, we examine the taming of fire, development of cooking, and the evolution of stone tool cultures. Fire made sleeping on the ground safer, whereas cooking made digestion easier. Stone tool cultures tell us about the tool makers and the pace of innovation.

In Chapter 36, we examine the sexual nature of humans and the unique physiological changes that makes us different from our ape cousins. Sexual selection appears to play a huge role in establishing the physiology of men and women and the drive to find the best mate.

CHAPTER 31

BIPEDAL WALKING

CONTENTS

31.1   SCOPE

Man is unique amongst the mammals in many ways. One of them is the combination of upright stature and bipedal walking. We can walk and run on two legs. Meanwhile, our nearest relative, the chimpanzee, scampers around on four limbs. What caused our apelike ancestors to adopt this unique method of getting around and what kinds of physiological adaptions did it require? These are the questions we will examine in this chapter. Scientifically, it is not a resolved issue. There are still differing opinions on it.

31.2   BIPEDAL WALKING PRECEDED BRAIN DEVELOPMENT

31.2.1   THE OLD VIEW OF THE MISSING LINK

In the early days of paleoanthropology, there were far fewer hominin species discoveries than there are today. Consequently, it was a lot harder to construct an accurate picture of our ancestral evolution. The speculation began when Charles Darwin suggested that man had evolved from the great apes and most likely from the great apes of Africa. This stimulated a lot of talk in the scientific community about a missing link between apes and man and what he might look like. The consensus was that the missing link would be physically and mentally intermediate between ape and modern man. In other words, that missing link would probably walk bipedally and would have a brain size intermediate between an ape and a modern human. Nowadays, we have discovered a multitude of hominin species and are gaining an ever-increasing knowledge of them. That knowledge includes more accurate dating of the fossils, determination of the climate at the time when they lived, knowing which plants and animals lived at that time, and better scientific tools to aid in interpretation of the discoveries. It also includes a good estimation of their ability to walk upright and a measurement of their brain capacity. Even a lone skull can tell us whether they stood upright based on the position of the foramen magnum. One thing we have learned is that walking predated bigger brains. Thus, our initial conceptual image of the missing link between ape and man was not quite accurate.

31.2.2 WE NOW KNOW THAT BIPEDAL WALKING PREDATED BIGGER BRAINS

It turns out that bipedal walking came long before the first appearance of bigger brains and evidence of stone tool making. In fact, our earliest link to tree-dwelling apes is the so-called man-apes. These man-apes, also called Australopiths, were essentially apes that had adapted to a life where upright walking on the ground was routine. As far back as 3.7 million years ago, man-apes had evolved a body suited to proficient upright walking. We know this because they left clear footprints imprinted by feet very similar to ours. These famous footprints were discovered by Mary Leakey in 1976 at Laetoli, a site in Tanzania. The footprints most likely were left by *Australopithecus afarensis*, whose skeletons showed significant bipedal adaptation but whose jaw shape and brain capacity compared more suitably with that of modern chimps. However, bipedal walking went back farther than Lucy's species. *Ardipithecus ramidus* is the name given to a 4.4-million-years-old hominid species who had an ape-sized brain and walked bipedally. These finds indicate that bipedalism definitely preceded brain enlargement.

31.3   SKELETAL AND BODY CONSIDERATIONS

An ape's body is ideal for swinging from tree to tree but is unsuited for bipedal walking. For one thing, its legs are too short and its arms are needlessly long. By contrast, the human body is ideally suited to bipedal walking and even running. Our longer, more powerful legs make us more efficient for walking and running. Moreover, longer legs require less up and down movement while running, thus saving energy. An ape's feet are specialized for grasping branches, whereas our feet are longer, have an arch, have shorter toes, and our big toe is in line with the other toes. Our feet are so specialized for walking that we have completely lost our ability to grasp branches with them. The ape's pelvic bones are not ideally shaped for walking and supporting the weight of the intestines. A more bowl-shaped pelvis developed in the bipedal hominins. Apes do not need to support their organs with their pelvis and consequently have broader hips.

The hole in the bottom of the skull, where the spine enters, is called the foramen magnum. It is positioned too far back in apes to best balance the weight of the skull if they were to assume an erect stance. The foramen magnum is more centrally located in bipedal walkers. So it is clear that extensive changes have occurred in our transition from tree-adapted apes to bipedal apes and to *Homo erectus* and finally to modern humans (Figure 31.1).

31.4   ELAINE MORGAN AND BIPEDAL WALKING

One theory of how we became bipedal walkers is told in a book titled *The Scars of Evolution – What Our Bodies Tell Us About Human Origins*. The author is the late Elaine Morgan (1920–2013). I knew her to the extent of trading emails with her for a number of years. She was a Welsh science writer and the author of six technical books. Four of those books were

FIGURE 31.1   Ape skeletons compared to human (Left to right: Gibbon, Orangutan, Chimpanzee, Gorilla, Human) (By Huxley, Thomas Henry, 1825–1895 (https://archive.org/details/evidenceastomans00huxl) [Public domain], via Wikimedia Commons).

related to the idea that our early ancestors had an aquatic environment, which led to our unique physiology. She was honored as one of the 50 greatest Welsh men and women of all time.

One reason that you might enjoy her book is that besides being easy to read, it contains an excellent analysis of the human speciation problem. Elaine analyzed the process of becoming a bipedal walker, what anatomical changes our bodies have undergone, and what problems we still suffer from our new skeleton. Knee-problems and lower back pain are two of the most common.

Morgan's book also addresses three other major issues of human evolution: (i) why our brains tripled in size from the time we were apes, (ii) why we lost our fur, and (iii) why we developed the ability to verbally converse. These three topics will be discussed in subsequent chapters, but Morgan explains all of them with a single theory.

31.4.1   THE AQUATIC APE HYPOTHESIS

Elaine Morgan spent much of her life pondering these questions and had a hypothesis for why our evolutionary journey made these changes in us. She believed that we were once a group of apes, who became isolated from the greater community by rising water, and whose survival depended on adapting to a life in water. Many thousands of years later, when changing conditions released us from our isolation, we had acquired an upright posture, an altered skeleton adapted for bipedal walking, loss of fur, and other important changes. She originally called her concept "the aquatic ape" theory. Later, she renamed it the "aquatic ape hypothesis."

Where might this isolation have occurred? The Great Rift Region near the Red Sea has been subjected to periodic flooding in some areas and draining in others due to the rifting forces, which has been operating on a geological time scale of millions of years. Morgan proposed that our ape ancestors might have become marooned there during one of the flooding periods. The Danakil Alps is one possible area. Our isolated apes may have adapted to an area that was neither sea nor land but a swamp or Everglades-type environment. We know that apes are capable of adapting to swamp living, because it has happened before. The extinct ape, *Oreopithecus*, lived such an existence. In fact, he was also called the Swamp Ape. Excellent fossils were preserved in hardened mud and exhibited some human characteristics, including bone structure adapted to bipedal walking. *Oreopithecus* is not believed to be a human ancestor but seems to have been headed that way before he went extinct.

31.4.2   APES ARE UPRIGHT WHEN THEY WADE IN WATER

Morgan believed that the switch to bipedal walking was unlikely to have originated in the forest or savannah but happened as a necessary adaptation for a swamp existence. One must keep one's head up to breathe in deeper water, which would make a quadruped stance impossible. She also argued that the loss of our fur is uniquely explained by having lived in water. Wet fur is useless for insulating our bodies from the cold. Aquatic mammals, such as whales, manatees, porpoises, and hippos, have also lost their fur and now depend on a layer of blubber beneath their skin for maintaining their body temperature. Fur is the best insulation in air, but blubber is the best insulation in water. Those aquatic mammals (e.g., otters, seals) that have fur have evolved a very specialized fur. It is oily and fine enough to trap air bubbles within it.

31.4.3   HUMAN BABIES ARE PLUMP

Human bodies do have an excess of fat cells compared with apes. In fact, humans have 10 times the number of fat cells of other animals. In general, the animals with numerous fat cells are either hibernating or aqueous. Humans do not hibernate. Human babies are born naked and plump with fat, whereas ape babies are born furry, but very thin. These differences suggest human babies needed blubber to insulate their hairless bodies against heat loss in the water. Ape babies were thin because they didn't live in water and their fur did adequately protect them against cold nights. Another advantage for fat human babies is their ability to float. They are at the water's surface where they can breathe. One downside to using fat to insulate our bodies is the inability of cut wounds to properly heal. The wounds tend to remain open and need stitches to close them. That is exactly our situation today.

31.4.4   MARINE ANIMALS EXPEL EXCESS SALT

Morgan also examined the glandular system of today's humans and apes. She argued that many of the changes that have occurred during our evolution make sense if we had experienced an aqueous adaptation phase in our past. As an example, humans use eccrine glands for sweating, whereas apes do not. This eccrine system may have evolved as a mechanism for expelling excess salt. Animals that live in saltwater tend to evolve salt-expelling mechanisms. They need to expel salt that is accidently swallowed, and different animals have evolved different methods to accomplish it. The marine iguanas of the Galapago Islands sneeze to expel salt. Further evidence of a marine history is that humans are missing a warning mechanism against the danger of low-salt content. Most animals know when they are low on salt; we do not. Other mammals seek out salt licks when their sodium is low. We may have lost our warning system for low salt when living in salty water, where the warning was unnecessary. This flaw occasionally becomes fatal for some of us..

31.4.5   CONSCIOUS BREATH CONTROL

Morgan also examined the uniqueness of human breathing. Unlike apes, our larynx (i.e., top of the windpipe) has descended to be at level with the esophagus. This descent does not happen in human babies until they are at least three months old. Early on, they are able to suckle and breathe at the same time. After descent, that capability is lost. Humans, unlike apes, have a velum, which can be raised or lowered to isolate the nasal passages from the mouth cavity. Morgan concludes that these features had evolved to enhance our survival in a watery existence. We also have control over our breathing unlike apes. For example, we can hold our breath for a specified time. This ability is what makes speech easy for humans and too difficult for apes.

31.4.6   LOSS OF ESTRUS AND BIGGER BRAINS

Other adaptive human features are a hymen and thick labia major in females, presumably to seal this area from water-borne things getting in. Loss of estrus in human females is another feature. Estrus is an evolutionary device to alert males both visually and through odor that a female has come in season and wants to mate. These signals would be undetected in water. An aqueous life also tends to facilitate the development of bigger brains. Seafood diets tend to produce a 1:1 ratio of Omega-3 and Omega-6 fatty acids, which is ideal for brain growth.

31.4.7   MISSING DNA SEQUENCES IN HUMANS

The aquatic ape hypothesis is mainly challenged by the Savannah hypothesis, which argues that the evolutionary changes (i.e., bipedal walking, loss of hair, speech, etc.) came about as a group of particular apes adapted to this new environment with wide open spaces between the trees. Morgan debunked their underlying assumptions in her book, with a factual argument comes from a genetic study of baboons. A type C retrovirus has altered the DNA in baboons, and over time they have become immune to the harmful effects of the virus. Yet other mammals could still be harmed by it. Examination of the DNA of the other African animals found specific DNA sequences that protected them from the virus. All African apes had the marker, but humans did not have it. The conclusion is that either humans did not evolve in Africa or they were isolated somewhere in Africa, where the virus couldn't reach them. The latter explanation seems more probable.

Whether you buy into Morgan's aqueous ape hypothesis or not, her book is valuable for an essential understanding of the physiological process of our evolution into a species so different from our closest relative, the chimpanzee.

31.5   OWEN LOVEJOY AND BIPEDAL WALKING

Owen Lovejoy is an anthropologist with a PhD from the University of Massachusetts. He is a prolific author, having published well over 100 articles. He has been active in many aspects of paleoanthropology including reconstruction of Lucy and a biological analysis of Ardi. In 1980, Owen Lovejoy argued that bipedal locomotion was an unlikely adaptation. It results in a loss of maneuverability, less stability, and less speed than quadrupedal locomotion. Man is the only mammal that walks upright on two legs comfortably and his body has undergone considerable change to make it so. Our feet can no longer grasp branches but are specialized for walking and running. Our pelvis has undergone extensive revision and our femur meets our tibia at an angle rather than in line like an ape. Lovejoy's explanation for this most unlikely adaptation is that bipedal walking allowed these particular apes to have a higher birth rate. He thinks that a poorly developed bipedal walker had no chance of survival on the open savannah, so the adaptation fully evolved in the forests beforehand.

31.5.1 APES ARE HEADED FOR EXTINCTION

All of the existing apes on our planet seem to be headed for extinction. Human interference in their lives is the visible culprit, but if humans had never existed, they would still be headed for extinction. The reason is that their reproductive strategy is at the extreme end of the spectrum. If we consider reproduction in general, we can argue the premise better: One extreme strategy (Type r) is many offspring but no offspring care. For example, trees dispense thousands of seeds in the hope that a few might survive. Fish lay hundreds of eggs, the male fertilizes them, and a few make it to maturity in order to reproduce and keep the species alive. The strategy opposite of this is to have few offspring but give them care, protection, and nourishment. This approach is called Type K.

Apes use an extreme Type K reproductive strategy. A single offspring is nurtured for approximately five years. The female does not come into heat (i.e., estrus) again until then. This system worked successfully when the forests dominated the landscape, but once the ideal environment degraded due to the Ice Age global cooling, survival became a harder task and the birth rate was barely keeping pace with the increasing death rate. Individuals were not being replaced at a fast enough rate for the species to be self-perpetuating when lethally stressed. Lovejoy argues that the bipedal apes broke out of this ill-fated reproductive strategy and found a better way

31.5.2   HUMAN PAIR BONDING AND CHILD CARE

The bipedal apes used a different reproductive strategy than tree-dwelling apes. That new strategy had the advantage of allowing the bipedal apes to have babies more frequently and yet still be able to care for them. Just as bonobos have a different culture than chimps, so these man-apes developed a different culture than the arboreal apes. The key difference was that males participated in the raising of the offspring by providing food and protection, whereas male apes indiscriminately mated with a female in estrus and contributed little to childcare. In the chimp and bonobo world, the father could be any one of the many males who had sex with the female in heat.

Lovejoy believes the pair-bonded man-ape couples had exclusive sex. This commitment was strengthened by the female being sexually available to her mate not only at estrus but nearly always. Eventually, estrus disappeared. It was no longer needed to summon males. Sex was practiced by the bonded pairs habitually. The disappearance of estrus strengthened the pair bond culture because the females were no longer signaling their sexual receptiveness to the group. With two infants at once, the need to walk upright and use both arms for carrying became a necessity. The female is thought to have spent her time in a safe place, whereas the male ventured out and brought food home.

Lovejoy's hypothesis seems impossible to prove or disprove. Most human cultures today have pair-bonded couples, but it seems a reach to think the custom began five or more million years ago.

31.6   CLIVE FINLAYSON AND BIPEDALISM

31.6.1   THE COMMON ANCESTOR WAS NOT A KNUCKLE-WALKER

Clive Finlayson was born in Gibraltar and is of Scottish descent. He studied zoology at the University of Liverpool in 1976 and got his DPhil from Orel College, Oxford. Professor Finlayson is a zoologist, paleoanthropologist, and paleontologist. His research on Neanderthals is discussed in his book *Humans Who Went Extinct.*

Reading Clive Finlayson's opinion on bipedalism helped me dispel a notion that had been troubling me. That troubling notion was that the required skeletal changes for a knuckle-walking ape to evolve into an erect-standing, two-legged walker seemed very formidable. The knuckle-walking chimps and gorillas get around just fine as they are. It is hard to imagine any circumstance where they would bear the discomfort of an erect posture to travel when they could knuckle-walk to the new location much faster (although walking in deep water is one exception where they would need to stand erect). Finlayson suggests that the common ancestor of all apes had an upright posture naturally and it was the chimps and gorillas that specialized by becoming quadrupedal knuckle-walkers. He bases this opinion on the fact that orangutans adopt an upright posture routinely as they transverse branches. Orangutans are known to be the first existing apes to split off from the ancient ape population, and therefore, they may be more representative of the common ancestor of apes. So it seems more believable that the bipedal apes never went through a knuckle-walking phase at all but evolved from an already erect-standing ape.

31.6.2 OPEN SPACES BETWEEN TREES KEPT EXPANDING

Clive Finlayson is not unique in saying that climate was important in human evolution, but he stresses its importance to bipedalism more than anyone else. He believed that Ramidus (see Table 31.1, which is the source of these names) was the first of the proto-humans to have to adapt to environmental change. He had to deal with a mosaic of woods and open grassland, whereas his ancestors lived in the forest environment to which they were adapted. Lake Man was challenged even more as the open areas were even more extensive. The time devoted to bipedal walking was increasing as the climate became drier and the open spaces increased. Finlayson says that only the specially adapted survived, whereas many perished as a result of a changing environment. Lucy and her kin adapted to even more open spaces than Lake Man. Walking became even more essential as the distance between trees increased. They still retained the ability to climb trees as indicated by their long arms and curved fingers. However, trees mainly served as refuges, sources of fruit, and places to spy from.

The line of descent from the small-brained, proto-humans to the Homo line is murky and controversial. In *Homo erectus*, we see someone recognizable as human. They had a bigger brain, shorter arms, longer legs than the australopiths, and had fully adapted to living on land rather than

TABLE 31.1 Proto-Humans and Their Environment

Name	Species Name	Time and Place	Environment
Ramidus	*Ardipithecus ramidus*	4.4 Mya, Ethiopia	Grassy woodland
Lake Man	*Australopithecus anamensis*	4.2 Mya, Ethiopia	Open woodland
Lucy	*Australopithecus afarensis*	3.9 -3.3 Mya, Ethiopia	Mainly grassland
Lake Rudolf Man	*Homo rudolfensis*	2.6 Mya, Kenya	Savanna
Erectus	*Homo erectus*	1.8 Mya, Kenya	Savanna

in trees. Lake Rudolf Man is one of the possible links in the evolutionary sequence. Although it is entirely possible that new fossil discoveries will present an even more suitable candidate.

31.7   DISCUSSION OF THESE COMPETING EXPLANATIONS

I think more paleoanthropologists lean more towards Lovejoy's theory than Morgan's. Many avoid discussing the topic altogether. Some of them object to the aquatic ape hypothesis as unproven and perhaps unable to be proven. Yet those same objections apply to Lovejoy's theory. We know that humans did become pair-bonded, lose estrus, and develop a culture around it because we humans are like that today, and our written history confirms it. What we don't know is whether that cultural change was concurrent with bipedal development or very recently or sometime in between. Similarly, Morgan argued that unique physiological traits in humans today all arose from our aquatic ape phase. Indeed, the strongest aspect of her theory is that so many different traits are explained by it. By comparison, Lovejoy only explains bipedal development and loss of estrus. One compelling feature of Lovejoy's theory is the superior birth rate. In the pages ahead, we will learn that these new creatures, the man-apes, were highly successful not only in surviving, but also in adapting to new and more dangerous environments, in evolving into many new species, and in leaving fossils to prove it over millions of years. It is puzzling that they did so well considering that they lacked speed to escape and lacked fangs or claws to put up a fight. We are hard pressed to explain it, but after Lovejoy, we have superior reproduction rate as one of the advantages. The arguments of Finlayson seem especially persuasive. The climate did continually change to a cooler, drier climate over millions of years, which gradually reduced the tropical forests, increased the open woodlands, and eventually produced deserts and savannahs. Bipedalism would have become a more and more useful method of locomotion under those circumstances.

31.7.1   PERHAPS A BIPEDAL COMMON ANCESTOR

Earlier in the book, I noted that Adrienne Zihlman proposed that our common ancestor with chimps was more like the bonobo than like the common chimp in body and social traits. She saw the bonobo as the most similar to early Australopithecines. It is also noteworthy that although both common chimp and bonobo are knuckle-walkers, the bonobo is more likely to use bipedal walking for locomotion than the common chimp. It very well may be that the common ancestor of man and chimp was even more inclined to walk bipedally than the bonobo is today.

Remember that Finlayson concluded that the common ancestor of the apes was of upright posture and that knuckle-walking was an adaptation that occurred with those particular apes that became ever more specialized as tree-dwellers. Chimps and gorillas are the knuckle-walkers amongst the living apes.

Assuming that our common ancestor with chimps and gorillas was not yet adapted to knuckle-walking, then it would not have been that big of a challenge for the early hominins to become exclusively bipedal walkers when on the ground. Bipedal walking may have been the most efficient way to get around on the ground at that time. However, to consolidate the bipedal specialization and account for all of the skeletal changes to improve bipedal locomotion, some kind of isolation of the man-apes seems vital. Without isolation from the original tree-dwelling ape population, it seems impossible to evolve into such a radically different species. Finlayson suggests that those proto-humans who adapted to the ever more open spaces had survival advantages over those who did not. Perhaps the isolation came as bipedal apes moved into the open spaces whereas the highly tree-adapted apes clung to the diminishing areas of tropical forests.

31.8   SUMMARY

It was thought back in Darwin's time that bigger brains and ability to walk bipedally must have developed together. However, as hominin fossils were discovered and dated, it was learned that bipedal walking evolved far earlier than bigger brains. Three different explanations for the origin of upright bipedal walking are discussed.

Elaine Morgan's aquatic ape hypothesis not only explains bipedalism but also loss of fur, sweat glands, and other human traits. Focusing on bipedalism, which is this chapter's theme, Morgan believes that common ancestor apes became isolated by geological events and were forced to live an aquatic life. An upright posture was necessary to have one's head above the water to breathe. When the opportunity to leave the confinement arose many generations later, bipedal walking was comfortable and natural.

The Owen Lovejoy hypothesis assumes that certain apes adopted monogamous male-female pairings, where the male participated in the feeding and rearing of the offspring. This reproductive strategy resulted in more viable offspring than for the chimp-like social structure observed today. Bipedal walking evolved from a need to carry babies, food, etc.

The Clive Finlayson hypothesis puts an emphasis on changing climate. Tropical forests were disappearing and being replaced by open woodlands. He believes that our ancestral ape relatives had not yet committed to knuckle-walking and practiced upright walking. Consequently, they could use bipedal walking to traverse between groups of trees without major discomfort and inconvenience. These more ground-mobile apes would have had an advantage over tree-restricted apes in finding food in a time of changing plant life. Their survival and reproductive success would lead to better walking ability over generations as skeletal changes made that possible.

CHAPTER 32

HAIRLESSNESS

CONTENTS

32.1 SCOPE

When we are watching our chimpanzee cousins at the zoo, we see similarities with ourselves, but we see big differences too. We are upright, taller, smarter, etc., but most obvious of all is that they are covered with thick fur while we are virtually hairless. If we had a common ancestor with fur-covered apes a few million years ago as DNA evidence tells us, then why did we lose almost all of that fur? And when do we think that hair loss occurred in the human evolutionary story. It might have been very difficult for me to provide a scientific answer to these questions, if I hadn't come across an article in the *Scientific American* magazine written by Nina Jablonski. She had analyzed the problem and arrived at promising conclusions. Consequently, this chapter is based mainly on her article.

32.2   WHY DO MOST MAMMALS HAVE FUR?

Mammals, unlike reptiles, are warm-blooded animals and must eat regularly to maintain their body temperatures or they will die. So the fur surrounding mammalian bodies helps them preserve heat. We need food to provide the energy necessary to maintain our body temperature. One benefit to preserving heat is that one can live longer on less food and increase the period between meals. Birds accomplish the same goal with feathers. In fact, today's birds are descended from feathered dinosaurs. The dinosaurs evolved feathers for warmth, not for flight.

In addition to fur's insulation value, it protects from abrasion, parasites, excessive sunlight, rain, and also serves as adornment. Having a fur coat is vitally important if you are a small animal like a mouse. The ratio of skin area to volume is unfavorable to small animals, who want to conserve their heat energy. On the other extreme, very large animals have an unfavorable skin area to volume ratio for losing excess heat and are in danger of overheating. Having a fur coat becomes a disadvantage if the large mammal lives in a tropical environment. Illustrating this principle is the fact that elephants, hippos, and rhinos are hairless in their African domains.

32.3   LOSING HAIR TO COOL OUR BODIES

Nina Jablonski is an American anthropologist and paleobiologist. She received her PhD in anthropology from the University of Washington in 1981. Professor Jablonski suggests that there was a time in our evolutionary history when it was vital to shed our fur in order to cool our bodies and, in particular, cool our brains. The driver was climate change affecting our African environment over the last few million years. The Ice Age with its cyclic nature had been converting a forested landscape ideal for the life of a tree-dwelling ape into a savannah-like landscape inhospitable to tree-dwelling apes. The bipedal apes of this time had to travel overland for longer and longer distances to find suitable food. Strenuous walking and running for long periods will raise one's temperature to a life-threatening level. Our main device for cooling is to sweat. Unlike apes, who sweat an oily fluid that coats their hairs, our ancestors evolved a better sweating device, namely an abundant of eccrine glands. Eccrine glands produce a salty, watery fluid, which evaporates and thereby cools us. These eccrine glands are more abundant than the ape's sebaceous and apocrine glands, and they are closer to the skin's surface.

32.4   THE AQUATIC APE HYPOTHESIS

Competing with Nina Jablonski's theory is the "Aquatic Ape Hypothesis" (AAH) originally postulated by Sir Alastair Hardy and argued for in several books by the late Elaine Morgan. Essentially, the hypothesis suggests that climatic conditions isolated a group of chimp-like apes millions of years ago where they had to spend most of their time in water. Upright posture was mandatory to keep one's head above water, and hair disappeared as it had for many aquatic animals (e.g., whales, porpoises, sea cows, hippos, etc.). A layer of fat grew to provide thermal protection.

Jablonski dismisses the AAH with three arguments, which I now paraphrase.

1. Aquatic mammals adapt to an aqueous life in a variety of different ways. Shedding one's fur is not a mandatory result of such adaptation.
2. Fossil evidence indicates crocodiles and hippos were numerous in the rivers and lakes at this period and would have been too dangerous to allow an aquatic existence of hominids.
3. The AAH is too complex, involving life in water and then readapting to life on land. The climate cooling theory is simpler.

32.5   THE AUSTRALOPITHICINES DID HAVE FUR

Jablonski further disagrees with the idea that the australopiths were hairless. Here is her reasoning: the early Australopithicines had small brains, long arms, short legs, and hands specialized for tree-climbing. They lived at a time when the continuous forests were changing into open woodlands.

They nested in trees at night like chimpanzees do today. When on the ground, they maintained a vigil and were always prepared to climb up a tree at the first sign of predators. The Australopithicines were a long way from the life style of *Homo erectus*, who used their long legs to run down game until it was exhausted and overheated. The Australopithecines did not have anything to gain by shedding their fur if heat loss from running was the driver. They would have been already furless if they had Morgan's aquatic ape phase behind them.

The global climate change continued to change the environment from forest to open woodland, which were becoming ever more arid and tree-less. Open woodlands changed into savannahs and deserts. The hominins had to adapt to savannah conditions, where trees were scare and often unsuitable for climbing. The evolution of the Homo lineage was one of those adaptations. Although who the earliest members of the Homo lineage were is still being debated, the *Homo erectus* species is well-defined. We discuss them next.

32.6 *HOMO ERECTUS* LOST HIS AND HER FUR COATS

Some anthropologists prefer to call the African *Homo erectus* by its own species name, *Homo ergaster*. However, I prefer to stick with the *Homo erectus* name. It turns out that we know quite a bit about this hunter-gatherer and that is because he left fossils of himself on three continents. Moreover, a very complete fossil skeleton was found and recovered by Richard Leakey's team near Lake Turkana. The Turkana Boy, as he has been called, had a body much like ours in size and proportion. Unlike the smaller Australopithicines, *Homo erectus* had evolved long legs and a tall, lanky body. This kind of body shape is of advantage in hot climates because there is more skin area per unit volume to facilitate evaporation of sweat. Analysis of his leg bones tells us he was capable of running game down.

His brain size was smaller than ours; 800 cc versus 1350 cc. Yet his brain was huge compared to the 400 cc brains of chimps and the early ape-men. Erectus had a very long existence compared to many species. It was somewhere between one and a half million and two million years. The later fossils had bigger brains than the earlier ones.

Here is the crux of Jablonski's hypothesis: Modern humans have a very high concentration of eccrine glands compared to other primates. This condition of sweat gland enhancement may have been well underway with *Homo erectus* 1.6 million years ago. Body hair became superfluous and actually interfered with sweating as a cooling process. Further adaptations to the exposed skin occurred to provide protection that the fur had previously provided such as abrasion resistance, sun-blocking, etc. For example, melanin-darkened skin evolved. The skin had previously been pink-colored as it is in chimps today.

32.7   HEAD AND PUBIC HAIR

Obviously, we didn't lose all of our hair. We have hair on our heads, armpits, and pubic area. Moreover, men have more hair than women generally and can grow beards and mustaches. Hair in the armpits and groin provide lubrication to an area where chafing might occur, and some suggest that it served to propagate pheromones. Hair on the head remained to shield the head from the intense heat of the sun. Kinky hair may be the best kind of hair for a hot climate because it provides a matrix where air can circulate to aid in evaporative cooling. The "Afro" hairstyle embraces these concepts. A thick layer of hair shields the head, but its texture allows air to cool the layer.

As for the differences in hairiness between men and women, sexual selection may be largely responsible for the way things are. Later in the book, we will see analysis that suggests that sexual selection in early humans was very important to our evolutionary outcome, but it also seems to operate both ways. That is, females are selecting males as is true for most species, but males are also selecting females, which is more unusual. Perhaps this is due to the culture of monogamous pairing, where males actually make a commitment to his mate and their children to support and protect them. So less facial and body hair but long head hair on females may have been selected for by men. Hairier bodies and faces on males may have been selected for by women. Our concepts of what makes an attractive mate are hard-wired in our genes. How we currently appear is the result of thousands of generations of humans making sexual selections for their mates.

32.8 MY TAKE ON THE HAIRLESSNESS THEORIES

Nina Jablonski made a strong case in her explanation of why we became hairless. Although antelope, deer, and other grazers can easily outrun us for a while, if we persevere in following them, we can run them down until they are helpless. Their brains overheat whereas ours do not. Humans have evolved ways of shedding the heat that builds up from prolonged physical activity. Think of the marathon runners of today. It is an incredible feat to be able to run over 26 miles in a matter of a couple of hours. Obviously, considerable evolutionary changes were needed to convert a tree-dwelling chimp-like creature into this remarkable running human animal in the span of a few million years. Dean Falk, in her book, *Braindance*, examined the evolutionary adaptations to the hominin brain to provide better cooling. Apparently, many parts of our ancestor's body were adapting to life on the savannah and all that entails.

Personally, I am not as dismissive of Morgan's aquatic ape hypothesis (AAH) as Jablonski has been. Morgan had devoted much of her adult life advocating for the AAH through a string of books, while striving for some acknowledgment and approval from professional anthropologists.

So, let's examine Jablonski's arguments against the AAH.

1. *Hair loss not necessarily an aquatic adaptation.* Yes, but it often is (whales, hippos, etc.) and it could have been for the aquatic hominids. Let's not forget the fact that newborn human babies are plump, hold their breath, and can swim in the water. That suggests an aquatic adaptation. The plumpness in babies is a human innovation; primate babies are skinny.

2. *Crocodiles and hippos would have killed the aquatic apes.* Landmass shifts and flooding have been one of the consequences of the rifting process in East Africa. It is possible that an elevated land area became isolated by invading seawater. The isolated hominids might have needed to get most of their food from the ocean. The expanded eccrine glands might have been an adaptation to expel excess salt from the body. No crocodiles or hippos would be in this scenario. If we apply Jablonski's argument about deadly predators to life in the savannah, hominins would not theoretically survive

there either because lions, hyenas, leopards, and other predators would have killed them all. Yet, they did live there and even thrive there.

3. *The AAH is too complex, requiring first aquatic living then adapting again to land.* True, it is more complex than the climate change theory, but that doesn't make it wrong. New species formation almost always involves isolation of a group of individuals and their progeny in order for it to happen. Isolation is harder to envision with the climate change theory whereas isolation is essential to the AAH. Furthermore, hominins have probably always lived close to water. Whether that might be a swamp, river, lake or ocean, there are better survival chances in being near a source of water.

The bottom line is that both theories have great merit.

32.9 SUMMARY

Nina Jablonski's theory builds from the fact that modern humans have a very high concentration of eccrine glands compared to other primates. These sweat glands may have evolved in *Homo erectus* as early as 1.6 million years ago. The need to cool the body had become vital for them. Body hair disappeared because it interfered with sweating as a cooling process. Melanin-darkened skin evolved from the pink-colored skin seen in chimps today.

The AAH by Elaine Morgan is an alternate explanation for our evolutionary loss of fur. Isolated apes had been forced into a life in water due to some geological event. Rifting in Eastern Africa has been creating such environments for millions of years. Our aquatic ancestor's fur disappeared as it had for many aquatic animals (e.g., whales, porpoises, sea cows, hippos, etc.) A layer of fat grew to provide thermal protection in place of the fur.

Jablonski dismisses the AAH with three arguments: (i) all aquatic mammals do not shed their fur, (ii) crocodiles and hippos were numerous in the rivers and lakes and would not have allowed an aquatic existence of hominids, and (iii) AAH is too complex of an explanation.

Jablonski's arguments do not nullify the AAH. Some aquatic mammals do shed their fur as the aquatic apes may have done, crocodiles are rarely found in salt water, which the aquatic environment may have been, and one additional step in a process doesn't disqualify it. The bottom line is that there is something valuable to learn about human evolution by examining each of these two hypotheses.

Finally, a third influence in our appearance is sexual selection. Perhaps much of the reason for our varying degrees of body hair, facial hair, and pubic hair has to do with sexual attraction.

CHAPTER 33

BIG BRAIN DEVELOPMENT

CONTENTS

33.1   SCOPE

One of the most interesting problems in human evolution is the question of why human brains grew so big and what evolutionary factors were responsible for making it happen. The Australopithecines underwent a remarkable process of becoming skillful bipedal walkers. Yet, over the two or three million years of their existence, their brain size changed very little. When the Homo lineage developed some two million years ago, their brain size underwent an incredible expansion where it tripled in size. Humans then had the greatest brain-to-body size ratio of any creature on Earth. How did it happen and why did it happen? We will explore these questions in this chapter.

33.2   HOMO BRAINS GREW BIGGER AND BIGGER

Our human ancestors, the Homo lineage, tripled their brain size over two million or so years of evolution. The Australopiths had not increased their brain size much in their millions of years of existence. However, once the Homo lineage began, brain size increased and continued to increase, as shown in Table 33.1. This was a costly change in terms of the energy required to maintain and run the brain. Our big brains require 25% of the energy we take in as nutrients and 20% of the oxygen that we breathe in. The fact that the Homo linage is associated with stone tools and weapons and evidence of tool marks on animal bones tell us that meat was a part of their diet. Meat eating and bigger brains seem to go together. Meat may have provided the additional nutritional energy to allow brains to grow bigger for the first time in hominin history.

The brain size increase was offset somewhat by a sizable reduction of our guts. The Australopiths had big guts as we can tell from their flaring ribcages. A reconstruction of the 3.2-million-years-old Lucy skeleton shows that her ribs flared outward from top of the ribcage downward. She and her kind (i.e., *Australopithecus afarensis*) had big guts. Now consider the modern ribcage; it is straight up and down. *Homo erectus* had a ribcage more like that of ours today. So over time, the length of the human gut decreased as the size of our brains increased.

One could make a generalization that the brains of hominins grew larger with time, and the data in Table 33.1 seems to confirm it. There is also a corollary that bigger brains correlate with greater intelligence.

TABLE 33.1   Cranial Capacity Increases Over Time

Hominin Species	When, years ago	Cranial capacity, cc
Chimpanzees	Today	350 to 400
Australopithicines	4 to 2 million	About 450
Homo habilis	2.5 to 3 million	About 612
Early *Homo erectus*	2 to 1 million	About 870
Late *Homo erectus*	1 million to 50,000	About 950
Homo heidelbergensis	800,000 to 200,000	About 1200
Homo sapiens	200,000 to today	About 1400

There are exceptions to this rule, however. For example, the species *Homo florensiensis* had very small brains; one individual had a cranial capacity of 380 cc. This brain capacity falls within the range of modern chimps, yet stone tools and evidence of cooking were found in the vicinity of the hominid remnants, indicating higher intelligence. These "Hobbits," per the nickname which caught on, lived on Flores Island in Indonesia. Seven skeletons were found as soft bone, not yet fossilized.

Another anomaly exists with the small skulls of *Homo naledi* recently discovered in South Africa. The cranial capacity of the males was 560 cc. and for the females 465 cc. This is markedly smaller than found for *Homo erectus* (i.e., about 870 cc.). *Homo naledi* has been recently dated to have lived around 300,000 years ago. It seems that these small-brained hominins transported their dead down dark and dangerous cave tunnels in a ritualistic burial. This is totally unexpected behavior for this brain size. *Homo florensiensis* fossils have been dated at 50,000 to 100,000 years ago (Figure 33.1). Sometimes island-dwelling species evolve toward smaller versions of themselves, but it is surprising that a hominin with a 380-cc brain would be able to hunt and make stone tools.

FIGURE 33.1   Figure 33.1 *Homo florensiensia* (https://upload.wikimedia.org/
wikipedia/commons/thumb/1/10/Homo_floresiensis.jpg/220px-Homo_floresiensis.
jpg). Homo florensiensia https://upload.wikimedia.org/wikipedia/commons/thumb/1/10/
Homo_floresiensis.jpg/220px-Homo_floresiensis.jpg.

Clearly, there is more to learn about the relationship between brain size and intelligent behavior.

33.3   BIG BRAINS AND THE HOMO LINEAGE

Paleoanthropologists work hard to draw relationship threads between the different hominin species. Arranging the different hominin species in chronological order helps us see the probable ancestral paths. Knowing the geographical range of the species also helps. However, comparison of morphological characteristics is complex and painstaking work. Paleoanthropologists may sometimes disagree with each other about proper species classification, but probably all of them would agree that the *Australopithecines* dominated the earlier bipedal times and *Homo*, the genus that led to us, developed from one of these australopithecine species around 2.5 million years ago. Moreover, the main criterion for being considered in the Homo genus has been possession of a big brain.

Big brains come with a high price, and that makes us ask why did evolution favor them. Part of the price is the high cost of maintaining them. Another cost of the big brain is burdening our species with having to care for helpless infants for two years or more. When the fetal brain grew to the point where it could no longer pass through the birth canal, babies arrived with only a partially developed brain. The remaining brain growth had to happen outside the womb. In our struggle for survival, the big brain must have served an important function, but what was it? We will now examine various explanations that have been offered for big brain development in our ancestors.

33.4   TOOLMAKING AND SOCIAL SKILL THEORIES

Imagine two groups of hunter-gatherers during the time of *Homo erectus*. Group A has better stone toolmakers in it than does Group B. The men of Group A can fashion sharper, more effective meat-cutting tools than those of Group B. All else being equal, we expect the population of Group A to increase faster than for Group B due to their more successful scavenging missions. It is also likely that the population of group B decreases due to its failed scavenging missions. The brain regions controlling planning ability, hand-eye coordination, and execution of Group A men have importance in survival, and the better genes for these regions get passed forward, whereas less successful Group B toolmakers do not. This culling process has been proposed as a mechanism driving bigger brains.

Another theory has to do with social skills in hunter-gatherer groups. The idea is that being able to read the intentions of others and to be able to influence others is a life or death skill set. Form the right alliances and you prosper. Form the wrong alliances and you may die. Some believe that these kinds of survival pressures created the need for big brains.

33.5   THE BIG BABY THEORY

This Big Baby Theory is discussed in Chapter 11 of the book *Big Brain* by Gary Lynch and Richard Granger. They begin by rejecting the idea that *Homo erectus* acquired a big brain due to an evolutionary stress such as the need for better toolmaking skills or the need for better social skills. Their reasoning is based on the fact that human brains have the same distribution of zones (i.e., mid-brain, diencephalon, neocortex) of the brain as that of any primate, whereas an increase in the percentage of neocortex relative to total brain mass would have occurred if toolmaking or socialization skills had improved due to stress-driven adaptation.

The authors suggest that babies became bigger simply as a by-product of changes to the skeleton as *Homo erectus* evolved into a better walker and runner. He developed longer legs, smaller gut, taller frame, and other adaptations. There was more room in the mid-section of the female's body for a larger placenta and larger babies resulted. Although the neocortex was in the same proportion as it would be in other primates, it was bigger as a result of being part of a bigger brain. So the authors believe that *Homo erectus* did not develop a bigger brain in order to be a better toolmaker. He became a better toolmaker because he had a bigger brain. He did not develop a bigger brain to become more socially astute. He became more socially astute because he had a bigger brain.

The theory has appeal because it eliminates the need to explain what kind of evolutionary pressure could have possibly driven the development of bigger brains. However, if bigger babies developed occasionally because there was more space available for the placenta to expand, it is easy to see how these bigger-brained hominins thrived and multiplied. *Homo erectus* lived in a very challenging environment and had abandoned the safety of tree-nesting that the Australopithecines had relied upon. He needed to be smarter and more socially connected to survive.

33.6   BIGGER BRAINS DUE TO SEXUAL SELECTION

33.6.1   *GETTING THE BEST GENES INTO THE NEXT GENERATION*

Many biologists have commented on the fact that our big brains are unnecessary for survival and that smaller brains would work quite well instead. In fact, our amazing brain expansion resembles a runaway phenomenon. There is a well-known mechanism that leads to runaway traits, and it is called sexual selection. The peacock's large and ornate tail is often cited as the classical example of a runaway sexual selection trait. So what is sexual selection? Charles Darwin pioneered the concept of sexual selection and cited numerous examples of it in the animal kingdom. Remember that offspring obtain half their genes from the father and the other half from the mother. In many species, females choose among a field of competing males with the goal of selecting the male with the best genes. The survival of the offspring is enhanced by selecting the most fit male to father them. When we observe sexual selection at work in the animal kingdom, its manifestations often seem to make the male less safe in his environment. For example, the peacock's large and ornate tail may be attractive to peahens, but it makes peacocks less camouflaged and therefore more vulnerable to predators. Sexual selection traits usually come with a price.

33.6.2   *THE CASE FOR AND AGAINST RUNAWAY SEXUAL SELECTION*

Geoffrey Miller, in his 433-pages book *The Mating Mind*, methodically explores the premise that sexual selection is the cause of big brains in humans. First, he examines the idea that the increase in the size of the human brain may have been due to runaway sexual selection. The extraordinary threefold growth of the human brain is unequaled in the animal kingdom. It turns out that many other extraordinary animal traits have been driven by runaway sexual selection. Perhaps human brain increase is also due to the same evolutionary pressure.

However, Miller is troubled by the fact that (i) the brain expansion should have happened in a much shorter time frame, (ii) it would have had to be driven by female mate selection, and (iii) polygyny should have been the societal norm. In other words, those males exhibiting the biggest brains should have been fathering most of the offspring as a result of females choosing them. In fact, these three conditions cast doubt on runaway sexual selection being the only cause of big brains because (i) the time span for brain growth was about one million years, which is far too long, (ii) monogamy, not polygyny, is the normal human situation, and (iii) human females obviously compete with other women for the best males when long-term relationships are sought. Thus, male selection is at work here too. Reproduction is essential for sexual selection to work, and reproduction primarily occurs in long-term relationships. In other words, runaway sexual selection was not the driver for big human brains. If it had been the driver, it would have happened in a fraction of the time, a few big-brained males would have been impregnating most of the females, and males would not have a choice in selecting their long-term partner. Other scientists believe that sexual selection is the dominant factor in human traits being what they are.

33.7   MUTUAL SEXUAL SELECTION COULD HAVE DRIVEN BRAIN SIZE EXPANSION

33.7.1   SEX FOR HUMANS IS DIFFERENT THAN SEX FOR APES

There are some important reproductive differences between humans and apes. Female apes physically display their sexual receptiveness during the most fertile time in their cycles, and coitus almost always results in pregnancy. Human females, on the other hand, evolved to conceal when they are most fertile. They engage in coitus throughout their cycle, and because the typical time that coitus occurs is not their most fertile time (i.e., ovulation), they do not get pregnant. The result was an unlikelihood of getting pregnant from sexual trysts and a high likelihood of getting pregnant during committed long-term relationships.

33.7.2   HUMANS AND SEXUAL SELECTION

Human females, just like all female animals, have more investment in reproduction than do human males. Females, not males, had the wombs that carried the fetus to term, and females produced the milk that fed the baby during its first few years. Females have more reasons to be selective about who the father of her children would be than males have in being selective in the mother of their children. Males benefit from having multiple partners if fathering many offspring is the goal. Females, who get burdened with pregnancy and childcare, have fewer chances than males to bring children into the world and therefore benefit from selecting the fittest male to father those children. The human male, unlike male apes, tended to form committed relationships with a female mate, so he also had a reason to select the best mate. His sperm would be solely invested in this selected woman for a multiyear period. Her fertility, fitness, and likelihood of being a capable and committed mother to his offspring were important.

33.7.3   FITNESS INDICATORS

The name of the game for selecting a mate for both males and females was to recognize fitness indicators. Miller defined a fitness indicator as a biological trait that evolved specifically to advertise an animal's fitness. He further identified the fitness indicators as ornaments, such as the peacock tail, and courtship rituals. Finally, he argued that the human mind evolved many of its most distinctive features during the human sexual selection process as fitness indicators. He substituted the phrase "healthy brain theory" in place of the abandoned phrase "runaway brain theory." He had a viable theory for why our big brains developed over time. *It was due to sexual selection, but one based upon both males and females selecting their mates through a screening process where fitness indicators governed their selection.* Moreover, the brain and its activities were the logical source of most fitness indicators. Behaviors such as language and art are products that could only come from a costly, complex brain. These behaviors advertise our fitness to the opposite sex. This theory not only explains why our brains grew bigger, but also explains why such nonessential activities as art, music, jewelry, gossip, and song became so important to our species.

33.8 WHICH THEORY IS BEST?

Some argue that big brains evolved due to natural selection, few argue that social interactions demand better brains, few others argue from the concept of sexual selection, and still some others argue they just happened as a result of the skeletal changes associated with bipedal walking. Whichever way they evolved, there was a heavy cost associated with the benefit of big brains. For one, human childbirth became a more difficult and dangerous experience. Moreover, the prolonged phase where the infant is helpless and requires care as the brain continues growing outside the womb is burdensome for the parents. It may be that all four theoretical causes of big brains may be partially true. Perhaps natural selection drivers such as hand-eye coordination in flint toolmaking were important in brain development. Perhaps social skills had life or death consequences and shaped brains, Perhaps larger wombs were possible due to skeletal rearrangements and bigger babies having bigger brains were born and thrived. And perhaps sexual selection led to brains larger than necessary for survival yet necessary for display of fitness indicators such as music, art, speech, and romantic thought.

33.9 SUMMARY

The average brain capacity of the different hominin species is one of the important features of interest to paleoanthropologists because we humans are the biggest brained creature accounting for size, because brain size seems to correlate with intelligent behavior, and because of something very unusual that happened during *Homo erectus* times. The australopiths never grew their brains much even though they had a few million years to do it. However, that all changed with the evolution of the Homo lineage. There seems to be runaway brain growth, where the brain capacity tripled compared with the australopiths brains. Human brains in proportion to body size are the biggest of any other species in the world.

This evolution of big brains by the Homo lineage is an important puzzle for science because big brains came with a big price tag. Big brains are expensive to maintain, and human babies are a burden for several years due to their helplessness. The addition of meat to the Homo lineage diet helps explain the source of additional energy, but the question of why big brains developed is still open for discussion. The following explanations were offered.

1. Stone toolmaking requires special mental skills. The brain evolved to advance these mental abilities.
2 Understanding fellow humans is a survival skill. The brains of people evolved to read other humans and thus stay alive and prosper.
3 Bigger brains resulted from physiological changes in the skeleton. Big brains developed because there was more room for them in the *Homo erectus* skeleton. New mental capabilities developed in the extra tissue.
4 Runaway sexual selection might explain the extraordinary changes. If a trait is part of mate selection, it can increase at a rapid rate.
5 Mutual sexual selection seems to be operative in humans. Selecting the best mate is important in passing good genes to one's offspring and helping them reach adulthood, where they can reproduce.

All these factors may be important, but sexual selection is mentioned by many paleoanthropologists as the main driver of human traits.

CHAPTER 34

SPEECH AND LANGUAGE

CONTENTS

34.1 SCOPE

The ability to communicate by talking and listening is unique to human beings. In this chapter we shall discuss what science knows about it. We really do not yet know when speech was first acquired, but there is reason to believe it was being used during the time of *Homo erectus*. We shall see some of the things we do know about speech centers in the brain. These facts have been learned from medical studies arising from speech or comprehension problems. One powerfully useful finding is the existence of the FOXP2 gene. It will be discussed. Speech led to language, and there are many languages in the world. The Proto-Indo-European (P.I.E.) language is of particular interest to English language speakers and people from Europe because it was the original language from which those languages developed. We shall explore its origins. Finally, we shall consider memes, which are concepts that live in the minds of humans. Memes, like genes, seem to have a life of their own.

34.2 SPEECH IS UNIQUELY HUMAN

There are many different animal species on our planet, but only one of them has developed the capacity of making and understanding speech, and that is us. Even our closest relatives, the apes, are incapable of it. Speech is no trivial thing. It is probably the single mostthing that makes us different from other animals and capable of improving ourselves. So we have to wonder when in our evolutionary path from bipedal ape to modern human did we acquire speech. Anthropologists have speculated about it and come up with different times and scenarios, but the question is still unresolved. Now it seems logical to me that human speech must have evolved gradually over the last million years or even earlier. *Homo erectus* is associated during this period with stone toolmaking, the use of fire, hunting large animals, and sleeping on the ground rather than in the safety of trees. These activities seem to require some communication between group members to facilitate learning these skills or planning coordinated events. Furthermore, we know that the brain of *Homo erectus* was significantly larger than that of the earlier Australopiths. What would drive brain expansion better than even more complex speech development?

34.3 SPEECH IS NOW HARD-WIRED IN HUMANS

Those of you who have raised children or watched your young grandchildren develop have noticed how tots learn to walk and talk. These uniquely human abilities are rapidly acquired by infants whether we try to help the process along or not. Their earliest attempts to talk sound like aimless babbling, but they are listening closely to the adults as they talk to each other and they are absorbing it. Youngsters are capable of learning any language. This enhanced ability to learn to speak and comprehend a new language in infants has a limited duration and then it gradually diminishes as the child ages. When we are older, it is hard work to learn a new language, as many of us know.

I find the phenomena of an infant acquiring the ability to converse an awesome thing. There is nothing like it on the planet. Apes cannot learn to talk, but they have been trained to communicate in verb/noun sentences using sign language. Yet if we compare the learning rate of a human tot to the smartest ape, the tot is overwhelmingly smarter.

34.4 SPEECH CENTERS IN THE BRAIN

34.4.1 BROCA'S APHASIA

Much of what we know about the brain comes from medical studies of people with brain defects. This includes problems with speech centers in the brain. As an example, Broca's aphasia is the inability to use or understand all but the simplest of grammar. A person with this condition would have difficulty understanding who had died in the sentence, "the cougar was killed by the grizzly." The ability of understanding the importance of word order is processed in the Broca's area of the brain, which was damaged in the case of this individual (Figure 34.1).

34.4.2 WILLIAMS SYNDROME

Damage to Wernicke's area of the brain results in mental retardation and a tendency to speak in a rich but senseless stream of words. In the case of William's Syndrome, which is a genetic disorder, affected children have an elfin facial appearance and a cheerful demeanor, but they are mentally retarded and tend to chatter on in an endless stream of vivid words. They have the capacity to learn language quickly but what they say makes little sense.

The genetic defect has been identified as a deletion of about 26 genes on Chromosome 7. It occurs as often as one in 7500 births or as rarely as one in 20,000 births. Heart problems are also associated with Williams syndrome.

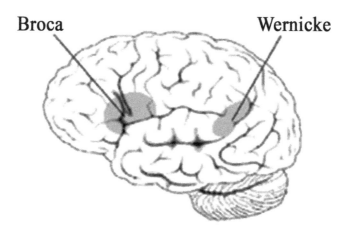

FIGURE 34.1   Broca's area (https://upload.wikimedia.org/wikipedia/commons/
thumb/0/03/BrocasAreaSmall.png/250px-BrocasAreaSmall.png). Broca's area https://
upload.wikimedia.org/wikipedia/commons/thumb/0/03/BrocasAreaSmall.png/250px-
BrocasAreaSmall.png.

34.4.3   SPECIFIC LANGUAGE IMPAIRMENT

Specific Language Impairment (SLI) is also a genetic disorder. There
is strong evidence that the condition is inherited. It is characterized by
reduced linguistic ability without a lowering of intelligence. There is
controversy regarding SLI where one school of thought focuses on sen-
sory impairment as being the reason for the problems, whereas the other
school believes that it is difficulty in understanding and using grammar
that is primarily the problem. The affected people have to store many
more individual words than normal people because they can't utilize
grammar rules effectively.

34.4.4   RESULTS OF NEUROIMAGING STUDIES

Neuroimaging experiments have shown that two neural networks exist
in the cerebral cortex that have to do with speech. One network ties
meaning to sounds, and the other network links the meaning of words
to their articulation. The study showed that various regions of the cortex

are interconnected. It was previously thought that speech was processed mainly in Wernicke's and Broca's areas of the cortex.

34.5   THE FOXP2 GENE AND SPEECH

Scientists have learned that a particular gene, called the FOXP2 gene, is vitally important to our ability to form words. That gene in humans has changed slightly from the corresponding gene found in apes. It codes for only two different amino acids than does the ape version, but that small difference may be critical. FOXP2 gives us the ability to control facial muscles as we form spoken words. People with a defective FOXP2 gene are severely handicapped in using language.

The most significant point is that FOXP2 is a regulatory gene. In other words, it orchestrates the expression of a whole set of other genes that make human bodies and minds capable of speech. More and more it seems that the difference between chimps and humans is not found in comparing our genes but in comparing the regulation of our genes. The mutation leading to our modern FOXP2 gene is thought to have occurred in the last 200,000 years. If so, the richness of our speaking and comprehension abilities may have increased markedly at this time. If that occurred, then the mutated FOXP2 gene would have been selected due to the advantages in communication that it would bring to its recipients.

34.5.1   HUMANIZED MICE

You might ask, "How sure are we that the FOXP2 gene really affects speech that much?" An experiment done by Wolfgang Enerd and team at the Max Planck Institute gives us some confidence that it really does affect speech and comprehension. They produced "humanized mice" by implanting the modern human FOXP2 gene in mice embryos. The altered mice exhibited very notable changes in their ultrasonic vocalizations compared to the normal calls when young mice are isolated from their mothers. The humanized mice also showed changes that have to do with articulation of speech and understanding of words. The conclusion was that the human

FOXP2 gene causes significant and extensive changes in the parts of the brain associated with language.

34.6   EVOLUTION OF LANGUAGE

Now we shall switch from discussing speech-related brain regions to the actual spread of languages themselves. Language is an interesting topic because it evolves just like biological beings evolve. The development of new languages is somewhat similar to the development of new species. When groups of humans get isolated from each other, their common language evolves differently for each of the isolated groups. Over many generations, the differences become so great that the groups are said to speak different languages.

Today, it is estimated that there are about 5000 different languages in the world. Admittedly, no one knows for sure when verbal communication took off, but we can surmise that our ancestors had enough verbal skills to coordinate the building of boats many tens of thousands of years ago. By 40,000 years ago, there is no doubt that we were verbally communicating. Cave paintings, bead jewelry, bone flutes, advanced stone toolmaking, and burials of the dead attest to it. We probably only had one major language at that time. As our ancestors migrated to virgin lands and became isolated into distinct groups, new languages evolved. Linguists have grouped languages into families and assumed that a proto-language existed for each family. The Australian continent had 270 different languages, of which more than half have gone extinct. This illustrates how complex linguistics can be. Nonetheless, tracking of language roots is another tool to be used in establishing human migrations.

You cannot read much about ancient human migrations without repeatedly seeing the name of Luigi Luca Cavelli-Sforza mentioned. He devoted a lifetime to advancing the state of knowledge in human genetics. Within his book *Genes, Peoples, and Languages* is a chart that correlates many similar languages of the world with the genetic similarities of their users. Moreover, he provides a detailed analysis of the data. Readers who want to learn more about this topic should read this book.

34.7   THE SPREAD OF THE PROTO INDO-EUROPEAN LANGUAGE

34.7.1   THE PROTO INDO-EUROPEAN LANGUAGE

The Indo-European language is important because nearly half of the world's population speak a form of it. Specifically, we are talking about a language that spread over Eurasia, the Americas, and Australia. The original language has evolved into different languages today. They include Spanish, English, Hindi, Portuguese, Bengali, Russian, Marathi, and French.

34.7.2   HORSEBACK RIDING AND THE SPREAD OF LANGUAGE

Horseback riding may have originated about 6000 years ago in the Ukraine. This predates the invention of the wheel and may have been just as important because horse-riding ability made the society very powerful compared to their neighbors who were still limited to walking. The horse rider's range of hunting and trade was vastly expanded, and their advantage in warfare was immense. Archeologists have devised a way of determining whether fossil horses were ridden or not. They look for wear marks on the fossil horse's teeth and see if they match similar marks on the teeth of modern horses wearing a bit. The Ukraine appears to be the original site of horseback riding and therefore may be the center for the proto-Indo-European language (P.I.E.). It seems likely that the spread of P.I.E. and horseback riding expansionists tie together. P.I.E. evolved into Sanskrit, Homeric Greek, Latin, French, English, Russian, Persian, and others.

34.8   MILK-DRINKING WARRIORS

Gregory Cochran and Henry Harpending don't deny that horseback riding may have been a factor in the spreading of the P.I.E. language but believe they have an even better explanation. They began their research by examining what had been deduced about the Proto-Indo-Europeans from an examination of the common words in P.I.E. It turns out that the P.I.E.

speakers were farmers who raised cattle, sheep, goats, and pigs. However, the cow was the animal that was central to their life and culture. The Proto-Indo-Europeans had wheeled carts pulled by oxen, weaved wool, and may had used bronze implements. They were war-like and raided cattle or sought revenge. They were divided into three classes identifiable by the color of their clothing: White for polytheistic priests, Red for warriors, and Black for herders and commoners.

Although an exact date is unknown, the P.I.E. language began spreading between 3000 BC and 2500 BC. The original site of the P.I.E. speakers may have been the grasslands between the Black Sea and the Caspian Sea, although this topic is controversial. Cochran and Harpending tend to look at genetic advantage in significant human events. The genetic advantage of the Proto-Indo-Europeans was lactose-tolerance. The normal condition is for adults to be lactose intolerant. Once infants are weaned, they no longer need the ability to digest milk. However, once those milk-producing animals had been domesticated, there was a huge survival advantage for the herders to be able to digest milk from those animals. The advantages that milk-drinkers had over non-milk-drinkers are: (i) that milk was available year-round, and (ii) it not only prevented starvation during lean times, but it built strong healthy bodies. Whereas the early grain farmers suffered from malnutrition and disease, the dairy farmers became healthier and more productive.

Dairy farmer/warriors living on extensive grasslands had another advantage over grain farmers, they were mobile. They could pick and choose their battle, and retreat indefinitely if necessary. Thus, Cochran and Harpending believe the spreading of the P.I.E. language was accomplished by lactose-tolerant cattlemen, who were healthier, produced more offspring, moved their food source with them, were good at conquest, and ruled as an elite.

34.9   MEMES

Richard Dawkins came up with a useful concept in regard to communication in human society. He had been thinking extensively about the replicating unit in biology, namely the gene. Successful genes increase in the population generation after generation. He realized that the connected brains of humans constitute a network wherein information can be sent, received, and stored. The unit of information that can be replicated he named the "Meme." It behaves somewhat like a gene, so he had meme rhyme with gene.

A meme might be a tune, a joke, a story, a recipe, or any other piece of information that can be replicated by communication between individuals. If the meme represents a catchy tune, a recipe for a delicious dish, a quotable quote, or other popular concept, it propagates itself by being repeatedly sung, recited, written, etc. Some memes are very powerful and affect human culture. For example, the concept of God is a meme. Reference to God can explain many questions that perplex humans at certain times in their lives. Why did my wife die? Why is my child blind? What happens to us when we die?

Memes can also evolve or mutate. Once the scientific revolution gathered momentum in the 17th century, it transformed civilization at an accelerated pace. Scientific memes such as the steam engine, railroad, automobiles, airplanes, and space ships swept around the world. Gunpowder, dynamite, tanks, and fighter planes caused wars to increase in severity. However, penicillin, antibiotics, sterilization, autopsies, and transplant surgeries revolutionized medicine. Memes are responsible for all we do as humans.

34.10 SUMMARY

The ability to communicate by talking and listening is unique to human beings. We are unsure of when it began, but *Homo erectus* engaged in activities that most likely required communication. They made stone tools, used fire, hunted, and traveled thousands of miles. They had much bigger brains than previous hominins, which could also indicate that they were communicating by speech.

Medical science has provided knowledge about speech and comprehension in the brain: Broca's aphasia is the inability to use or understand all but the simplest of grammar. This condition is associated with damage to Broca's area of the brain. In the case of Williams syndrome, children are mentally retarded and tend to chatter on in an endless stream of vivid words. The genetic defect has been identified as a deletion of about 26 genes from Chromosome 7. Specific Language Impairment is a genetic disorder characterized by reduced linguistic ability without lowering of intelligence. The affected people have to store many more individual words than normal people because they can't utilize grammar rules effectively.

The FOXP2 gene is vitally important to our ability to form words. The human version codes for two different amino acids than does the ape version. People with a defective FOXP2 gene are severely handicapped in using language because of an inability to control facial muscles. The important point is that FOXP2 is a regulatory gene that orchestrates the expression of a whole set of other genes that make human bodies and minds capable of speech. Regulatory genes are the key to understanding the differences in chimps and man. Our version of the modern FOXP2 gene may have occurred in the last 200,000 years. The proof that our mutated FOXP2 genes are special came when they were implanted in mice embryos. The humanized mice exhibited very notable changes in their ultrasonic vocalizations and other changes.

Speech led to language, and there are many of them in the world. The P.I.E. language is of particular interest to English language speakers and people from Europe because it was the original language from which those languages developed. The invention of horseback riding, which may have originated about 6000 years ago in the Ukraine, may be the mechanism by which P.I.E. originally spread throughout the world. The horseback riders had a big advantage in that their range of hunting and trade was vastly expanded, and their advantage in warfare was immense.

Cochran and Harpending have a different theory. They think it was the milk drinkers who spread the P.I.E. because they had the genetic advantage of lactose-tolerance. Milk was available year-round, milk prevented starvation during lean times, and milk built strong bodies. Whereas the early grain farmers suffered from malnutrition and disease, the dairy farmers became healthier and more productive.

The concept of memes is important to understanding human evolution. Memes are human ideas that live, mutate, and evolve in their environment, which are the connected minds of humanity. Memes, like genes, seem to have a life of their own.

CHAPTER 35

FIRE, COOKING, AND TOOLS

CONTENTS

35.1 SCOPE

One big difference between chimps and humans is our ability to use tools to shape the world in our favor. While it is true that chimps have used twigs to extract termites or rocks to break open nuts, they have done nothing like using fire for warmth, or cooking their food, or shaping selected rocks into cutting instruments. In this chapter, we will see how such actions helped us adapt to non-arboreal environments. Using these new tools also had the effect of changing us. We were becoming more human, and fire, cooking, and stone-cutting tools helped direct that evolutionary path.

35.2 THE BENEFITS OF TAMING FIRE

When early man learned to utilize fire in their daily lives, it gave them a tremendous advantage in their daily struggle to survive and prosper. Fire gave them protection against dangerous predators of the night. *Homo erectus* no longer relied on nests high in the trees for sanctuary, as did his Australopithecine ancestors. Predators have an advantage at night when it is dark, but their fear of fire protected our ancestors. It also gave them illumination, making the dark hours available to use for the first time. It also gave warmth, allowing early man to stave off cold at night or during cold spells. Fire may have been used in hunting. Terrified animals could be driven off a cliff. Finally, fire led to the practice of cooking food, which is discussed at length below (Figure 35.1).

35.2.1 WHEN DID MAN FIRST TAME FIRE?

There is a lot of evidence that mankind routinely used fire by 125,000 years ago. However, most anthropologists believe that early man had tamed fire long before that. *Homo erectus* appears to have used fire about 400,000 years ago based on charcoal deposits found in excavations. For example, unequivocal evidence of fire use has been found at Qesem Cave, near Tel Aviv. It consisted of ash, burned soil, and butchered bones of aurochs and horses.

FIGURE 35.1   Early man warming up by the campfire.

Recently, evidence of fire use at Wonderwerk Cave in South Africa pushes the date back to one million years. Sediments clinging to charred items had been heated, and this strongly suggests fires were used by the inhabitants of the site. Researchers also found hand axes and other stone tools. Perhaps *Homo erectus* had even conquered fire as early as 1.5 million years ago. Evidence from a *Homo erectus* site at Koobi Fora suggests it. East African sites show clay sediments undergoing color change that would be only possible with baking temperatures above 400° C. They are dated at 1.42 mya. The taming of fire and using it for cooking may even go back to 1.9 mya. The teeth of *Homo erectus* suggest it. Aside from the fossil site evidence, we can see the physical changes in the skeletons of the Homo lineage, which are best explained by the advantages gained from fire.

Homo erectus was also the first hominin to migrate out of Africa and settle in Europe and Asia. Having the use of fire would have facilitated such migrations. Now he was able to deal with colder climates by building fires to keep warm. Otherwise he would have to deal with frostbite, hypothermia, and other problems associated with the cold. If they indeed had fire during these migrations, it must have become part of their toolkit even earlier than 1.5 mya because they were in Europe by then. The development of speech may have also been furthered as these hunter-gatherers shared a campfire at night and told stories and shared experiences.

35.3   COOKED FOOD HAD AN EFFECT IN SHAPING THE HOMO LINEAGE

Richard Wrangham wrote his book *Catching Fire* that argues persuasively for the significant importance of cooked food in the evolution of man. He tells us that the invention and adoption of cooked food in our diets was a major game-changer. The high points of his arguments are presented below.

35.3.1   COMPARING MAN TO CARNIVORES

Carnivores eat almost nothing but meat, whereas humans cannot do that and survive. Wrangham tells us that too much protein acts as a poison in our systems, producing ammonia in the blood, liver and kidney damage, loss of appetite, dehydration, and finally death. Like carnivores, we have smaller guts than we would have as herbivores. However, there are some important differences: Carnivores process food in their stomachs much longer than we do, making them much better at digesting meat than we are.

35.3.2   EVOLUTIONARY CHANGES

Chimps comprise the species that is our nearest living relative. They live on mainly a vegetarian diet of fruits, leaves, and other fibrous foods. Such foods take a big gut in order to extract energy through digestion. Australopithecine fossil skeletons exhibit downwards-flaring ribcages, indicating they had big guts too. As the Homo lineage progressed over time, we see that the ribcage flaring to accommodate a massive gut disappeared and the slim waistline of a physically fit modern man appeared in its place. A modern man's gut is only 60% of what would be expected from a primate of equivalent size.

Chimps spend a lot more time chewing fibrous foods that humans do, and they are better equipped to do that chewing. They have bigger, sturdier jaws, bigger teeth, and large muscular lips. Humans have evolved to have much weaker chewing ability because cooked foods are so much more energetic and easier to digest. We no longer have large jaws, teeth, and chewing muscles, nor do we spend as much time chewing. Cooking was responsible for those physical changes in us.

35.3.3   INABILITY OF HUMANS TO SURVIVE ON RAW FOOD

Wrangham not only touts the benefits of cooked foods over raw foods but asserts that our bodies have so radically adapted to eating cooked foods that we can no longer survive on raw foods. He cites studies where individuals lost weight and became dangerously ill trying to survive on raw vegetation alone. Cooking has changed our digestive systems in an irreversible manner.

35.3.4   ENERGY ADVANTAGE TO COOKED FOODS

Although the U.S. government's National Nutrient Database claims no energy difference between raw and cooked foods, Wrangham has collected considerable data that says otherwise. Especially compelling are data on the energy gained from digesting food in the small intestine (i.e., ileal digestibility). He reports that starchy foods are the main ingredient in breads, cakes, and pasta. They are also in cereals from wheat and rice. Starch is present in 63% of the average diet. Studies of ileal digestibility show that most starchy foods are 95% digested if cooked but fall in the 48–71% range if eaten raw. The mechanism by which starch is made more digestible through cooking is gelatinization. Essentially, the granule structure of starch is broken down by heat and that makes it easier to digest.

He also examined the effect of cooking on proteins. Again, he found that cooked meats are more digestible and provide more energy than raw meats and other proteins. Contrary to the long-held belief that raw eggs are more nutritious than cooked eggs, the opposite was found to be true. In fact, almost twice as much energy is available in cooked eggs over raw eggs. Other studies show similar advantages with meat proteins. The mechanism is denaturing of protein structure via heat. The folded structure of proteins is broken down, making them easier to digest.

35.4   TOOLMAKING

35.4.1   THE ADVANTAGES OF STONE TOOLS

Stone toolmaking has been an evolutionary process as evidenced by the increasing sophistication of the industry over the two and one half million years since it began. The first stone tools were probably rocks, already naturally broken and leaving a cutting edge. Finding such tools was a search-and-selection process. If on occasion no suitable stones were found, early man may have attempted to break the cobblestones himself with variable results. The stones being struck with a hammerstone break with a conchoidal fracture, leaving a sharp edge. At some point, he learned that he could improve on nature and fashion the blade that he sought through a pioneered technique. The art was passed from father to son. This earliest stone tool culture is called the Oldowan culture and is associated with *Homo habilis*. Louis Leakey was a pioneer in uncovering its fossils in Old-uvai Gorge, Tanzania, where such stone tools were found in large quantity.

The Homo lineage is associated with a hunter-gatherer lifestyle in a savannah-type environment, in which meat began to play a major role in their diet. Most paleoanthropologists believe that hunting of game was not the original way meat was obtained and that scavenging was more typical. The key to successful scavenging was to sever chunks of meat and get to safety before predators, such as lions and hyenas, competed for the carcass. This scenario puts a premium on finding or making the sharpest cutting blades and axes possible. We know that these activities occurred because animal bone fossils have cut marks on them, which could only be made by man. Other stone tools, such as scrapers, were developed to prepare the animal pelts for tanning. Animal hides were used for clothing, bedding, and waterproof structures. Knapping, as stone toolmaking is called, is skill-intensive. Modern-day knappers have spent long hours developing the skills to make their first successful blade or spear point. Some scientists believe that the brains of early man evolved to master this skill over many generations.

35.4.2   STONE TOOL CULTURES

The different toolmaking cultures obtained their names from archeologists, who relied on the differences between tool and weapon relics to identify the dates and clans during excavations. I am not going to do justice to this tool culture topic in this small space. Yet, I wanted to give the reader an inkling of what it entails. Table 35.1 shows an overview of 18 different toolmaking cultures.

35.4.3   THE HOMO ERECTUS TOOL CULTURE

Homo erectus is associated with the Archulean tool culture (Figure 35.2). These tools are usually quite symmetrical and have a teardrop or oval appearance to them. The hand axes made by these ancient ancestors have been found in large quantities. The skill required to produce these tools indicates a mental leap compared to the Oldowan culture. Hand axes

TABLE 35.1   Chart of Stone Tool Cultures

Part of Paleolithic	Culture	Period
Lower	Oldowan	2.6–1.7 Ma
	Riwat	1.9–0.045 Ma
	Soarian	0.5–0.13 Ma
	Archulean	1.8–0.1 Ma
	Clactonian	0.3–0.2 Ma
Middle	Mousterian	600–40 ka
	Micoquien	130–70 ka
	Aterian	82 ka
Upper	Baradostian	36 ka
	Chatelperronian	41–39 ka
	Aurignacian	38–29 ka
	Gravettian	29–22 ka
	Solutrean	22–17 ka
	Magdelenian	17–12 ka
	Hamburg	14–11 ka
	Federmesser	14–13 ka
	Ahrensburg	12–11 ka
	Swiderian	11–8 ka

Source: Wikipedia https://en.wikipedia.org/wiki/Aurignacian.

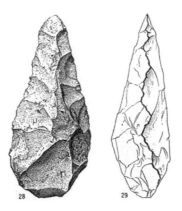

FIGURE 35.2   The Archulean hand axe (Courtesy of Wikimedia Commons). Archulean Hand ax (https://upload.wikimedia.org/wikipedia/commons/thumb/d/df/Characteristics_ boldly_flaked_Abbevillian_hand-axe_Wellcome_M0015180.jpg/512px-Characteristics_ boldly_flaked_Abbevillian_hand-axe_Wellcome_M0015180.jpg).

have been referred to as the Swiss army knife of these times because they were used for a variety of purposes. Some Archulean hand axes in Kenya, Africa, have been dated at 1.76 million years old.

35.4.4 *EUROPEAN STONE TOOL CULTURES*

Table 35.2 provides a little more detail on some important cultures that developed in Europe or Eurasia during the Ice Age. The data comes from Brain Fagan's book *Cro-Magnon* and may differ slightly from the previous table at times. The Mousterian culture is closely identified with the Neanderthals (Figure 35.3). These archaic humans were discussed in Chapter 13. They were strong, stocky ambush hunters who had adapted

TABLE 35.2 Six Major Stone Tool Cultures

Stone Tool Culture	Period	Comments
Mousterian	100 ka to 30 ka	Neanderthal societies, which were widely dispersed over Europe and the Near East.
		Predominately made from flint.
Chatelperronian	Circa 40 ka	An enigmatic cultural identity in France and Northern Spain.
		(Possibly both Neanderthal and Cro-Magnon)
		Distinctive flint knife with single cutting edge.
Aurignacian	39 ka to 29 ka	Widely distributed over Europe and into the Near East (Cro-Magnon) heavy use of bone and antler in toolmaking.
Gravettian	29 ka to 18 ka	Widely distributed over Europe with local variations Microliths, nets, and burins
Solutrean	18 ka to 17 ka	Northern Spain and Southwest France
Magdelenian	17 ka to 11 ka	Northern Spain to Central Europe. Decorated tools using engraved figures.

FIGURE 35.3   Mousterian stone tools (Courtesy of Wikimedia Commons).

physically to the frigid conditions of the time but were not as inventive as the Cro-Magnons occupying their territory. They lived predictable, unchanging lives over a couple of hundred thousand years. Unlike the Cro-Magnon, they did not work with antler and bone and did not experiment with different stone-working techniques. They gradually diminished in number and finally went extinct. The Cro-Magnons are featured in Chapter 14 and are responsible for the remaining stone tool cultures in the table. The different cultures reflect adaptation to the cyclic climate of the Ice Age and the types of game pursued. For example, during colder climates, flint and other toolmaking rocks were scarce materials. This led to conservative use of these materials by fashioning smaller points and other tools. These smaller tools are called microliths. Antlers and bones also became raw materials for tools and weapons.

The tool industry evolution also shows innovative ways of catching small game, birds and fish. Nets were an important invention in this regard. Barbed points were another.

For those of you interested in getting a deeper understanding of the stone tool culture differences, the book *The Dawn of Human Culture* by Richard Klein and Blake Edgar is worth reading. It is loaded with drawings of the many stone tool types and styles.

35.5   SUMMARY

With the origin of the Homo lineage, our human ancestors began to take control over their own destinies and made adaptations to the non-arboreal lifestyle in which they found themselves. The taming of fire was a huge advance, which provided them with protection from predators, warmth on cold nights, and a resource for cooking. This may have happened as far back as 1.9 mya. When man advanced into the higher and colder latitudes, fire was essential for survival. Cooking was another huge advance. It made food more digestible, killed harmful germs, and changed our metabolisms. Cooking is most likely responsible for our smaller teeth, jaws, and chewing muscles. It may also be responsible for our smaller gut.

Stone toolmaking allowed meat scavenging to be possible and allowed us to harvest all the different parts of an animal. Hides could be used for a variety of uses, including clothing. Warm clothing was essential for the Ice Age adaptation in Eurasia. The sophistication of toolmaking seems to increase throughout our prehistoric past. The Oldowan culture is the oldest and the crudest. The Archulean culture was more advanced but lasted for an immensely long period. In more recent times, humans became far more innovative and updated their stone toolmaking technology quite frequently.

CHAPTER 36

SEX

CONTENTS

36.1   SCOPE

We humans had a common ancestor with the African great apes (i.e., chimpanzees and bonobos) that lived some five to seven million years ago. That common ancestor was more like a chimp or bonobo than like a modern human. We humans have come a long way from our tree-dwelling life in a tropical forest. In this chapter, we will examine how we have changed sexually. Not surprisingly, sexual selection plays a major role in shaping our sexual nature. The process of selecting the fittest mate possible has driven our evolution and made us the human beings that we are. Sex and mate selection for humans, and somewhat for apes, are social phenomena. However, from an evolutionary viewpoint they are mainly reproductive phenomena. Successful reproductive strategies make the species more fit for survival and more populous. Humans have developed a more successful reproductive strategy than apes based upon comparisons in population growth. Humans are seven billion individuals strong whereas apes are on the edge of extinction. Keep that fact in mind as you read this chapter.

36.2 INSIGHT FROM THREE LIVING SPECIES

Humans have some sexual traits that are unique to the animal kingdom and are even markedly different from comparable traits in our great ape ancestors. The two ape species most similar to us humans are the common chimp and the bonobo. These apes are similar in appearance, but due to a million years of isolation from one another, have developed different sexual practices. Common chimps are a male-dominated society, where aggression is part of their culture. Bonobos have a less aggressive, female-dominated social culture, which relies on sex to reinforce relationships to a high degree. Not only is male-female sex frequent for bonobos, but female-female genital rubbing is common and male-male sex is observed too. Female apes have even more sexual activity when they ovulate because they advertise their condition with a prominent swelling of the genital area and a pink coloration. Many male apes answer the invitation. Human females, by contrast, do not have any indicators to announce their ovulation. In other words, for some reason or another, human females have evolved hidden ovulation. One possible explanation is that their upright posture hides their genital area more than is true for quadrupedal apes.

Humans are quite different from either of these ape relatives. Human women have evolved prominent breasts and buttocks, have hidden ovulation, experience difficult births, and end their reproductive years with menopause. Some observers believe that human facial lips have evolved to attract the opposite sex as well. Human lips are unique in the animal kingdom in their size, shape, and color. They are perhaps a visual reminder of the female vulva, but at any rate, they are used in love-making extensively. Human males have evolved muscular bodies and a longer, thicker, more flexible penis than do apes. Humans have evolved a different skeleton than apes as we adapted to an upright posture and bipedal locomotion. Coitus in humans is usually done in a face-to-face position whereas the quadrupedal apes usually do it in a rear entry position; that is, the male approaches and enters the female from behind. Humans and apes are all highly social animals, and sex is part of social bonding for all three. Grooming is the most common bonding activity among the fur-covered apes. Humans lost their fur over a million years ago, and conversation took over from grooming as the behavior used for bonding.

36.3   HOW DID THESE UNIQUE TRAITS DEVELOP?

Scientists always want to know how and why significant changes occur. We are especially curious when it comes to the sexual evolutionary history of our own species. Charles Darwin did not put his great mind to work on the problem until after 1859, but in 1871, he came up with the concept of sexual selection. When mating is restricted to selected members of a species, that process can have profound effects on the evolution of the species. Unfortunately, the power of his concept was not really appreciated until 100 years after he proposed it. Darwin believed that sexual selection was responsible for many of the differences between humans and apes and for most of the differences between humans themselves. Today, there are several evolutionary psychologists who are explaining how we have evolved in the unique manner that we have, and they are using the principles of sexual selection to do it. It is not enough to have superior survival traits. You must also have viable offspring to carry those superior genes into future generations. That is where sexual selection becomes important to how we evolve. I am relying on one of those evolutionary psychologists for much of the material in this chapter. He is Geoffrey Miller, and his book is *The Mating Mind*.

36.4   MONOGAMY, POLYGYNY, AND HAREMS

36.4.1   *SEXUAL DIMORPHISM IN APES AND HUMANS*

Sexual dimorphism has to do with the relative size of the male and female of a species. Among spiders, the female is usually the larger and more dangerous of the two. Among mammals, the male is usually bigger of the two. As a rule of thumb, the greater the size difference between male and female, the less the male involvement in child-rearing. The smaller the difference, the greater the male role is in child-rearing.

Amongst the ape species, sexual dimorphism varies. The male gorilla is much bigger than the female, and he may head up a harem of females. The male gorilla guards and protects his harem and offspring. Orangutans are also sexually dimorphic, but socially the male and female tend to live apart as individuals. The male orangutan contributes his sperm to the pairings, but that is the extent of his involvement in child-rearing. Chimp and bonobo males, on the other hand, are only slightly bigger than the females. Chimps and bonobos have a polygynous type of society, where male chimps line up when the female is in estrus.

Human males are slightly bigger than human females, and the typical social pattern is for males and females to form long-term bonded relationships. When we look at the hunter-gatherer societies, which still exist today, the tendency is not for life-long unions but for a series of committed unions during one's lifetime. Unlike the apes, human males do participate in child-rearing. The level of promiscuity in humans is far less than in chimps and bonobos but greater than that in gorilla families.

36.4.2 THE SPERM WARS

One of the ways that we can look into our distant unwritten history is through examination of the testicles of living apes and humans. It turns out that chimpanzees have the largest testicles of the great apes whereas those of gorillas are relatively small. The male gorilla had no incentive to grow larger testicles because he had sole domain over his harem of females. Chimps, on the other hand, had high incentive to grow bigger testicles. Their sperm had to compete with other sperm for the race to fertilize the female's egg. When female chimps come into heat, they tend to have coitus with every male in the group. It is up to the sperms to fight it out. You can visualize how over countless generations the males with the greatest quantity of ejaculate and the most potent swimmers would father the most offspring. Their offspring would compete in more and more difficult contests, and testicles would grow larger over generations.

Where are humans in this testicle size picture? We are intermediate between gorillas and chimps. We were not as polygynous as chimps, but we were not entirely monogamous either.

36.5 HIDDEN OVULATION

36.5.1 MENSTRUAL CYCLES AND ESTRUS CYCLES

Fertile women menstruate, which means they did not conceive during their last cycle, and consequently they are shedding the lining of their uterus. This lining is called the endometrium. Their bodies will construct a new endometrium in anticipation of a pregnancy in the next cycle. Menarche is when menstruation first begins. For human females, this is usually between 11 and 14 years of age. For the great apes it is age 8 to 10 in the wild, but age 6–7 if living in captivity. Menopause is when menstruation ends. Ovulation also ends, and the fertile time of the woman's life is ending. She may be in her 30s or even older than 50 when menopause begins. Although it is more difficult to measure, chimps and other primates seem to have a menopause too.

The menstrual process allows us to bring new human beings into existence and is an essential part of being human. However, menstruation is unique only to certain primates (e.g., monkeys and apes), bats, and the elephant shrew. All the other mammals have estrus cycles rather than menstrual cycles. One essential difference is that in estrus, the endometrium is not shed but is reabsorbed if conception does not occur. However, what is more important to the evolution of human culture is the limitations on sexual activity. Estrus cycle female animals are only sexually active when in estrus, whereas menstrual cycle female animals can be sexually active anytime. Actually, mammals can be categorized according to their estrous behavior. Monoestrous species are in heat once per year and include: bears, foxes, and wolves. Diestrous species (e.g., dogs) are in heat twice per year, and polyestrous species (e.g., cats, cows, and pigs) are in heat several times per year. Some polyestrous species (e.g., sheep, goats, deer, and elk) are in heat during fall and winter, whereas others (e.g., horses, hamsters and ferrets) are in heat during spring and summer. Humans, monkeys and apes have evolved a different social structure than these other animals in part because of the freedom to have sexual relationships at any time.

36.5.2   ADVERTISING ESTRUS

Primates, which menstruate, may also physically advertise that they are in heat and available for sex. They are said to exhibit estrus. Chimp females have swollen, colorful genitals during ovulation. It becomes a pink protuberance, which is sometimes quite large. Males are attracted and vie for her attention. The swelling lasts for about 10 days. Ovulation is midway between menstrual cycles. Bonobos have an extended estrus. They are in heat for 75% of their menstrual cycle. Orangutans, although lacking physical signs of ovulation, will approach the males and solicit sex. Female gorillas also approach the males when in heat.

36.5.3   HIDDEN OVULATION AND PREGNANCY

Human females do not advertise they are ovulating nor do they approach the males for sex because they are ovulating. At least, we are not aware of them doing it during historic times. This hidden ovulation eliminated the males-lining-up-for-sex phenomena that chimp societies experienced. This is an important aspect of human sexual evolution. It is impossible for male chimps to know if they are the actual fathers of any baby chimp. However, it is usually easy to tell who the father is in human societies. Committed male-female relationships are the key. Whereas brief sexual encounters had a poor likelihood of resulting in pregnancy, prolonged relationships would very likely result in pregnancy.

36.6   HUMAN SEXUAL SELECTION

Females of all species have a strong interest in selecting the fittest fathers for their offspring. Females, who are the sex carrying the eggs, tend to be more selective in choosing a mate than do males, who are the carriers of the sperms. The reason is that females have more invested in their eggs than males have in their sperms. Plus, carrying the child to term is also a large investment. On the other hand, males can best increase their genes into the next generation by mating with as many females as possible. Hence, they tend to be less selective in picking sex partners.

It is axiomatic to sexual selection theorists that it is the females of a species who do the mate selecting. This is obviously true for birds. The females retain the safety of dull camouflage coloring, whereas the males adopt bright decorative colors and adornments, sing serenade songs to lure the females, and even build and decorate nests to entice the females. So why in the human society are human females so much more attractive than human males? The answer is that males are selecting female mates at the same time that females are selecting males. In humans, it appears that both sexes are seeking the fittest, most attractive partner that they can get. That seems to be the human mating process. One gets the very best partner that one can get. If a male is at the top of the desirability list, he may be able to get the best female. However, if he is only middle of the list, then a middle-of-the-list female may be the best he can get. As generations of this type of sexual selection process go by, the species improve. Here is why: the offspring of the fittest survive and thrive, whereas the offspring of the least fit tend to die. The best genes increase in frequency and the worst genes are eliminated.

The human sexual experience was adding in something new. It was no longer strictly an act of procreation; it was romance. Feelings of curiosity, warmth, lust, awe, and adoration became an important part of the mating experience. Anyone,who has been in love knows what this is like. Better lovers had an evolutionary advantage.

36.6.1 THE ATTRACTIVE FEMALE

What makes a woman sexually attractive? Perhaps it is her youthfulness. This is an important fitness indicator to interested men. Genetically it is important because fertility wanes with age in females. Fertility wanes with age in men too but at a slower rate. How is youthfulness determined? Youth is evident in many ways: smooth skin, firm breasts, a spring in her steps, and a high energy level. However, a sexually attractive woman has more than youth going for her. She has traits that say she is sexy, but also say she is healthy and fit. Symmetrical features and absence of flaws have long been recognized as signs of beauty. Female breasts and posterior are prominent features for judging symmetry. The face is another. Is the right side a mirror image of the left? Full, kissable lips are sexy, but what about her eyes? The eyes are the window to the soul they say.

What of the genital area? Upright walking has changed this area to something quite different from what a chimp exhibits. The same parts are there (clitoris, labia, urethra, vagina), but a chimp has leathery patches to make sitting on tree limbs more comfortable, whereas human females hide the features of their vulva behind a triangular display of pubic hair. The clitoris is an organ, which is somewhat like a miniature penis, and which has as its sole purpose to provide pleasure to the woman. Moreover, it may play a role in the sexual selection of a male mate by measuring the skill level of her lovers.

Human females then have several indicators of their sexual attractiveness, not to mention their warm personalities and other mental characteristics. Thus, we can conclude that human males must have been actively selecting for these female traits for them to ever exist in the first place. It is doubtful that males were being that selective for cases of opportunity sex, so it is very likely that they were being selective for longer-term commitments. In today's societies, it is generally believed that lifelong commitments (i.e., marriages) are expected, children should be supported until they can make it on their own, and the male is usually the primary breadwinner of the household. So, today's male has good reason to be selective. In earlier times, the male commitments were far less stringent. Perhaps males learned to relish committed relationships. They had regular sex, have the children that they had fathered, and had a female to tend to them. They were selecting for more than regular sex; they were selecting for a lifestyle.

36.6.2 THE ATTRACTIVE MALE

What makes a human male sexually attractive? Hard-wired into all females is the need to screen prospective mates for maximum fitness. The usual indicators are symmetry, smooth skin, absence of flaws, energy level, strength, and intelligence. Thus, tall, muscular, handsome men are automatically candidates for further evaluation. For women, personality traits rank high too. Women do not have the muscular strength of men and are at a disadvantage if things turn violent. They want a high achiever, but one who can be generous, kind, and protective. She is highly alert to indicators for these attributes.

How about male genitals? Are women selecting for penis attributes? Modern surveys usually report that while males are stimulated by pictures of genitals, women are disinterested and not aroused. Maybe so, but the human penis is obviously a product of female sexual selection. If it wasn't, it would be short and pencil thin like those of our ape relatives. Instead, it is the largest penis of any primate, and it is thick and flexible. It is flexible because it is missing the bone found in ape penises. Most mammals have this bone, but horses and humans do not. In order to generate an erection, the penis becomes engorged with blood.

So human men have a special penis shaped by the evolutionary forces of sexual selection. How did that ever occur? Women don't even seem interested in looking at them. Although for much of our ancestral history, people were always naked. There is a better view of the penis in bipedal hominids than you get from a quadrupedal ape. When males get an erection, it is more eye-catching than when they are flaccid. While I pondered these facts and got nowhere, the best answer jumped out of Geoffrey Miller's book. Women select men who have penises that feel good inside their vaginas. This accounts for the size and shape of the male penis perfectly. Size and shape do vary somewhat in different parts of the world, but so do female vulvae.

Desmond Morris in his book, *The Naked Ape* even went into more detail explaining why a large penis might have evolved in humans. He believes as a general rule that female mammals do not experience an orgasm. Since he is a zoologist, I assume he is correct about this. So it is very noteworthy that human females certainly can and do experience an orgasm and that their pleasurable experiences are important to keeping human pair bonds relationships together. Morris reminds us that sex for humans is quite different than for mammals in that prolonged sex play precedes the act itself and that intercourse lasts far longer with humans than for other mammals. Sex is not done only for procreation in humans; it is also done for maintaining pair bonds. He argues that although the penis is not in direct contact with the female clitoris during coitus, it is big enough in girth to work the muscles around the clitoris and bring the female to a climax. He also notes that unlike quadrupedal animals, she would lose the sperm if she quickly assumes a vertical position after sex. However, if she needs to rest after having climaxed, sperm is retained longer and pregnancy is more likely. Thus, genes favoring offspring are passed on and those genes have thrived.

So the evolutionary story of the human penis might be something like this. Ancestral women routinely had sex with different men and that was the culture. Some men had penises that felt better and gave them more orgasms. They preferred to have sex with these better-endowed men. The men with whom they had sex more often were usually the ones who got them pregnant. Over many generations, pencil thin penises with bones disappeared, and larger, thicker, more flexible penises took their place.

36.6.3 THE SEXUAL BRAINS

Isn't a person different once you know them? We may be attracted to a person due to their appearance, but that becomes less important as we grow to know the inner person. Speech is unique to humans and it allows us to see that person inside the physical body. Sexual selection, humans talking to each other, and accelerated brain growth are all interrelated. Females had a lot at stake in selecting the right male mate because he might be a dangerous threat to her and her children. Strong, aggressive males were attractive, but they were also potentially dangerous. Our female ancestors of long ago relied on conversation to reveal the real person inside that attractive exterior. The genetic effect of this selection process over many generations was to change the typical male into a safer, more caring person. The crueler, more abusive males didn't get to father as many offspring.

Males soon learned the rules to this mating game and learned to hide their defects long enough to win over the damsel. It became a game of perception, cunning, and intelligence. There is also the game of each sex striving for the highest rated member of the opposite sex possible. So not only were males wooing high-rated females, females were wooing high-rated males. Conversation and courting behavior were tools to win the best mate possible. In addition to your physically attractive assets (or lack of them), you could sway judgments if you painted the right picture of yourself and acted accordingly. Brains became a sexual selection organ.

36.7 SUMMARY

We assume that our common ancestor with chimps must have been something like the chimps and bonobos that exist today. That common ancestor was a fur-covered, quadrupedal, tree-dwelling ape. We assume that they were as polygynous then as they are today. The female chimps in heat mate freely with all the males in the troop. Fossil evidence indicates that our direct ancestors evolved to become more like modern humans during the time of *Homo erectus*. A hunter-gatherer life style had developed, and larger brains, longer legs, smaller guts, and loss of hair occurred during this time. The power of sexual selection operating through the process of males choosing females and females choosing males had a transformation effect on the Homo lineage over many generations.

Our ancestors began to develop features designed to attract the opposite sex. In the case of females, breasts evolved to stay firm and full even when not lactating. A slim waist and shapely buttocks also developed to arouse the male suitors. She also began to hide all signs of when she was ovulating and most fertile. The social structure became less polygyamous and more monogamous, which was necessary to care for infants over a long dependent period. Male-female bonded pairs became the norm. Males attracted the female's eye when they were tall, muscular, and healthy looking. A larger, more flexible penis evolved as a result of the female selection process.

Despite their physical attractiveness, males with their greater strength could be a threat to females and their offspring. Sexual selection was directed to choosing males with both vigor and a tendency to kindness, patience, and forbearance. Intelligence, conversation, and behavior were indicators of a good male mate. These processes of humanization continue in *Homo sapiens* to this day.

PROBLEM SET FOR PART VI

QUESTIONS

Q1.   How does the aquatic ape hypothesis explain the switch from qua-
       drupedal to bipedal walking?

Q2.   Why do you think the apes are going extinct, whereas humans are
       over-populated?

Q3.   Today's chimps and gorillas are knuckle walkers. Their bodies are
       adapted to a quadrupedal locomotion on the ground. Finlayson
       suggests that the common ancestor of chimps and humans might
       have been bipedal. How would that difference affect the evolution
       of bipedal apes?

Q4.   Jablonski seems to have the better explanation for loss of fur if it
       happened to *Homo erectus*, whereas Morgan has the better expla-
       nation if it happened to the earliest Australopiths. Explain why that
       might be.

Q5.   Gary Lynch and Richard Granger did not believe that *Homo erec-
       tus* acquired a big brain due to an evolutionary stress such as the
       need for better toolmaking skills. What was their reasoning?

Q6.   Describe Geoffery Miller's "Healthy Brain Theory," which he
       believed explained the evolution of our big brains.

Q7.   How did the humanized mice experiment show that our human
       version of the FOXP2 gene gives us our speaking ability?

Q8.   The original horseback riders might have spread the P.I.E. language
       from the Ukraine. Present reasons why this might be so.

Q9.   We humans have a gut that is 60% smaller than predicted for a typi-
       cal primate of the same size. Explain why ours is so much smaller.

Q10.   Match the hominin with his associated stone tool culture: Homi-
       nins (Cro-Magnon, *Homo erectus, Homo habilis* and Neander-
       thal) Stone tool culture (Oldowan. Mousterian, Aurignacian, and
       Archulean).

Q11.   The degree of sexual dimorphism in a species is a useful tool for
       guessing about the male's involvement in parental care. What is
       sexual dimorphism and what might it tell us about male behavior?

Q12.   A male chimp can never be certain that any of the baby chimps in
       the troop carry his genes because the mother had sex with many
       males. It isn't in his interest genetically to contribute to the child
       care of the baby chimps because of this uncertainty. How is the
       human situation different?

Q13.   Human babies are helpless for at least two years whereas chimp
       babies become independent much sooner. Therefore, it is in the
       human mother's interest to select a mate who is supportive. Dis-
       cuss this from the female's viewpoint of a million years ago.

PART VII

HOMO SAPIENS DOMINATE

Something special happened to our ancestors around 50,000 years ago. They became a lot more innovative than they had been before and it worked quite well for them. Moreover, they started appreciating the non-essential things in life that make us human; and that make us different from all the animals. Some refer to this period as the Great Leap Forward. Suddenly, we see art, music, and spiritual appreciation. Numerous cave walls in France and Spain are decorated with polychromatic images of animals that lived as late as 40,000 years ago. Sculptures of fertility goddesses abound. Stone tool cultures advance at a record pace.

By 10,000 years ago, the Ice Ages were over and there was a resurgence of plants and animals. The hunter-gatherer life style had existed for perhaps over a million years, but agriculture and domestication of animals eliminated the threat of starvation and allowed populations to surge. The stationary farmer faced new problems including disease and malnutrition. The pastoralists (i.e., sheep and/or cattle ranchers) thrived and developed a warrior culture.

Finally, we reflect on what our investigation into human evolution revealed and speculate on what we see for our future.

CHAPTER 37

GREAT LEAP FORWARD

CONTENTS

37.1   SCOPE

An intellectual awakening occurred about 50,000 years ago in Europe and the Middle East in the period known as the Upper Paleolithic. After many millennia of an unchanging Stone Age culture, innovation became the order of the day. The anthropologist Jared Diamond gave a name to this innovative period. He called it the "Great Leap Forward." Our ancestors, the Cro-Magnons, seemed to hit upon a winning strategy; they used their inventive skills to adapt to the challenges they faced. We know this happened because they left behind evidence of their advancements. They filled their lives with better tools and weapons, built huts to live in, and sewed clothing, but most unexpected, they expressed themselves with art, music, and spiritual connections. Despite the harsh arctic climate of Ice Age Eurasia, these people thrived and made a great leap forward towards becoming modern humans. In this chapter, we explore this important time in the human story.

37.2 BACKGROUND

Long before our *Homo sapien* ancestors migrated out of Africa, another kind of humans had already occupied Europe and the Middle East. They were the Neanderthals, who acclimated to the region and its Ice Age climate over a period of 250,000 years. Our ancestors first encountered Neanderthals in what is now Israel at least 100,000 years ago. They would encounter them again as they migrated into the landmass of Eurasia beginning about 50,000 years ago. As we learned in Chapter 29, interbreeding between our ancestors and Neanderthals occurred at several times during our push northward.

The Neanderthals did not participate in the Great Leap Forward as far as can be determined. In other words, they did not exhibit the innovative spirit demonstrated by the Cro-Magnon people. In fact, our superiority in dealing with daily survival may be a dominant reason for why the Neanderthals went extinct some 30,000 to 40,000 years ago. We were killing the game animals they depended upon. It is ironic that our ancestors, who were adapted to warmer climates, should outcompete the Neanderthals, who were highly adapted to an Ice Age climate. It was our ancestors' willingness and ability to find innovative solutions to the new survival challenges that made the difference. The Great Leap Forward was a radical change for humanity. Innovation is an integral part of being human, and its origins appear vividly in Eurasia from 50,000 years ago.

Different authors report various dates for the beginning of the Great Leap Forward. Some say 35,000 years ago, which is the radio-carbon-dated age of the oldest caves in France. These caves contained paintings of animals, which are now extinct. However, the cave art is only a part of the story. The impetus of The Great Leap Forward resided in the spirit, intelligence, and attitude of the people migrating into Europe and else-where. That migration began something like 50,000 years ago. We have already traced that out-of-Africa migration in Chapter 28. Now we will get a look at the minds of the people themselves.

37.3   PROOF OF A WINNING STRATEGY

One sign of adaptive success for a species is an increasing population, and our migrating ancestors were very successful by this yardstick. An experiment was conducted by Henry Harpending and associates that proved that human populations were rapidly expanding during the Great Leap Forward era (see *The Journey of Man* by Spencer Wells, pp. 91–93). Harpending had collected mtDNA specimens from 25 worldwide popula-tions, and except for two of them, the specimens showed that exponential growth had occurred beginning 50,000 years ago. The population trends are determined by an analytical process called the mismatch distribution. Moreover, the populations in Africa, Asia, and Europe seemed to expand independently. The African expansion began 60,000 years ago, the Asian expansion 50,000 years ago, and the European expansion 30,000 years ago. Archeological evidence agrees with these findings.

37.4   FASTER RATE OF HUMAN EVOLVING

Gregory Cochran and Henry Harpending, in their book *The 10,000 Year Explosion,* argue that contrary to the pronouncements by some anthro-pologists that human evolution has ceased in modern times, the opposite is true. It has actually speeded up. They point to the sudden burst of inven-tiveness in Europe 30,000 years ago as extraordinary. The inventiveness of man burst forward suddenly against an archeological record of very slow and gradual change. They think there was a genetic reason for the unusual behavior. They think that rapid genetic change caused human evolution to speed up. Genetic change requires that mutations have occurred, but mutation rates are constant for all practical purposes. So what else could have introduced rapid genetic change? Their answer to that question is that it resulted from interbreeding with a separate species. They believe that the modern humans entering Europe some 40,000 years ago had inter-bred with the Neanderthals, and favorable genes spread quickly through the human population. They admit that they are speculating, but feel that the likelihood is high that they are correct. It seems they were unaware of Pääbo research, which proved that Neanderthal DNA is a small part of our DNA. On the other hand, I am unaware of Pääbo ever linking his discovery to the jump in inventiveness seen in the Great Leap Forward.

37.5   A NEW WAY OF MAKING TOOLS AND WEAPONS

We have already discussed the Aurignacions, Gravettians, Solutreans, and Magdalenians in Chapter 14. The names derive from the unique characteristics of their tool and weapon industry. Constant innovation produced new and better weapons more rapidly than ever before. Finding flint and other kinds of rock suitable for toolmaking was a much bigger challenge in Ice Age Europe than it had been in Africa. The Cro-Magnons solved that problem in two ways: (i) they used alternative materials, such as animal bones, mammoth tusks, and antlers for tool making, and (ii) they made smaller and smaller flint blades (i.e., microliths). Moreover, they were inventive in many ways. They fashioned sewing needles out of antlers and developed the skill set to sew hides together according to a preconceived plan. Now they could prepare snug-fitting clothes, boots, blankets, and tents. We doubt that the Neanderthals ever made this mental leap. Instead, they are thought to have draped animal pelts loosely over their bodies for warmth. The Cro-Magnon also developed the spear thrower, which allowed them to kill game animals at a distance from themselves. The spear thrower extended the throwing arm, allowing greater throwing distance and accuracy. The Neanderthals, lacking this technology, continued ambush hunting at close range and suffered the effects of the animals fighting back.

37.6   NEW METHODS OF ACQUIRING FOOD

The Cro-Magnon became better and better hunters with the result that many of their game animal populations were dwindling. The invention of the spear thrower (called an atlatl) was a game changer. By hurling lightweight spears and by generating greater penetrating force, they brought down horses, mammoths, bison, red deer, and other prey. However, during the colder cycles of the Ice Age, reindeer meat became a major part of the diet.

They also turned their attention to smaller, more populous prey, such as rabbits, hares, birds, and other small game. Snares and traps were contrived and smaller spear points became part of their tool kit. Oceans, rivers, and streams were now of increased interest as a source of food. Fish (salmon) and seals were hunted, and appropriate weapons were invented to be more effective. Harpoons with barbed points, fishhooks, and nets increased their yields.

37.7   PREHISTORIC CAVE ART

Perhaps, the discovery of prehistoric cave art is the most convincing evidence of a fast-evolving type of human, which had never been seen before. I mean evolving in the sense of a more active, intuitive, and curious mind. For it is in the cave art that we see artistic skill in capturing the essence of lions, horses, bison, ibex, reindeer, rhinos, owls, and others. Artists today would be challenged to do as well. The antiquity of these paintings is astounding; most of these animals went extinct long ago.

Consider the cave paintings discovered in France and Spain. Nearly 340 caves have now been found that contain prehistoric paintings on their walls and ceilings. Cave art has also been found in Asia. For example, on the Indonesian island of Sulawesi, the cave art depicting a pig has been dated to 35,400 years ago. Australia has cave art dating back to 28,000 years and some art depicts megafauna that have been extinct for 40,000 years. The earliest cave art in Europe was done in the Aurignacian period about 30,000–32,000 years ago. The Chauvet Cave in France contains beautiful paintings of animals in lifelike poses. Radiocarbon dating is ideal for establishing dates in this timespan.

Much of the prehistoric cave art depicts animals of the last Ice Age. These animals include bison, horses, aurochs, and deer. Although the animals are often exquisitely and realistically drawn, humans are symbolically drawn, usually shown as stick figures. Perhaps there was a taboo about drawing humans realistically. One common theme is hand outlines of humans on the cave walls. These were made by blowing pigment on a hand, which was held firmly to the cave wall.

37.8   EXAMPLES OF PREHISTORIC CAVE ART PAINTINGS

The four photos in Figure 37.1 typify the prehistoric art found in the caves of France and Spain. These paintings allow us to see into the minds of our ancestors who lived 30,000 years ago.

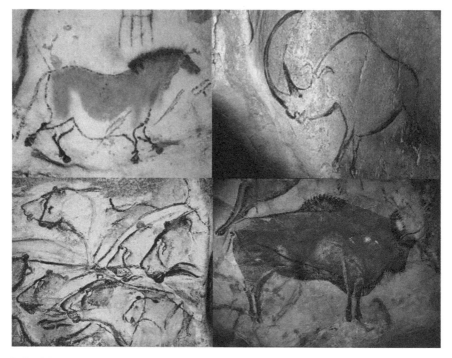

FIGURE 37.1 Prehistoric Cave Paintings of Europe (Upper right is painting of a horse from Lascaux Cave, Lower right are lions from Chauvet Cave. Upper left is a rhinoceros from Chauvet Cave, and lower left is a bison from Altamira Cave.) (Photos are courtesy of Wikimedia Commons). [Horse: https://upload.wikimedia.org/wikipedia/commons/0/07/Lascaux2.jpg; Lions: https://upload.wikimedia.org/wikipedia/commons/thumb/2/2d/Lions_painting%2C_Chauvet_Cave_%28museum_replica%29.jpg/512px-Lions_painting%2C_Chauvet_Cave_%28museum_replica%29.jpg; Rhino: https://upload.wikimedia.org/wikipedia/commons/thumb/6/6f/Rhinoc%C3%A9ros_grotte_Chauvet.jpg/512px-Rhinoc%C3%A9ros_grotte_Chauvet.jpg; Bison: https://upload.wikimedia.org/wikipedia/commons/thumb/c/cc/AltamiraBison.jpg/512px-AltamiraBison.jpg.].

37.8.1   MORE CAVE ART IMAGES

To view numerous cave art images and get a real appreciation for these ancient artists, I recommend the following website.
https://images.search.yahoo.com/yhs/search;_
ylt=A0LEVvRatmVXTVUAkoMnnIlQ;_ylu=X3oDMTByMj
B0aG5zBGNvbG8DYmYxBHBvcwMxBHZ0aWQDBHNlYw
NzYw--?p=Age+of+Prehistoric+Cave+Art+Paintings&fr=
yhs-mozilla-001&hspart=mozilla&hsimp=yhs-001

37.8.2   AN OVERVIEW OF CAVE ART

The caves containing prehistoric art are not normally open to the public, but there is a book that can give you a tour of some major caves. It is *Cave Art* by Jean Clottes. It is divided into four sections as per Table 37.1. This breakdown of time might be an appropriate way for us to think of cave art too.

The first European artists to paint images on cave walls were the Aurignacians. The amazing thing about their art is the advanced skill level displayed in it. A mastery of spatial perspective and motion is evident as well as creativity and an ability to render animals in their natural settings. Obviously, these artists had honed their skills long before attempting to decorate the walls of the Chauvet cave. Perhaps these ancestors of ours

TABLE 37.1   Organization of Jean Clottes' *Cave Art* Book

Section	Title	Years ago	Comments
1	The Age of Chauvet	35,000 to 22,000	Chauvet Cave
			Aurignacian and Gravettian art
2	The Age of Lascaux	22,000 to 17,000	Lascaux Cave
			Solutrean art
3	The End of the Ice Age	17,000 to 11,000	Niaux Cave
			Magdelenian art
4	After the Ice Age	After 11,000	Prehistoric art
			Various locations

practiced their skills on surfaces that were less permanent than the walls of a deep cave.

37.8.1   WHAT MOTIVATED THE ANCIENT ARTISTS TO PAINT ON CAVE WALLS?

Theories have come and gone concerning the motivation that led to cave art painting. These caves were maintained and visited over a 25,000-year span, so there was a special importance to these magical chambers. The most popular theory today is that the art was strongly connected to the shaman of these clans. The Cro-Magnon, with their inquisitive minds, may have questioned why do we have to die, what powers govern life and death, can we reach the spirit world and save our very sick child, and is there life after death. The shaman was the intermediary to the spirit world and may have used these cave paintings to create a sense of awe in the cave visitors.

37.9   SCULPTURE

Animals are also the most common themes in sculptures during the Paleolithic Age (Figure 37.2). Many have been found carved in ivory or bone. Venus figurines are also common to the time. These female figurines were carved from soft stone, bone, or ivory. Sometimes they were formed from clay and fired, and represent the oldest known ceramics. Over 140 figurines have been discovered, most of them from 26,000 to 21,000 years ago, although some date back to 35,000 years. The figurines have been commonly found in Europe but range across Eurasia all the way to Siberia; why they were carved is a mystery. Some speculate that they had religious importance or were erotic art, or were fertility goddesses.

37.10   JEWELRY

Shell beads have been associated with our ancestors well before they migrated out of Africa. They are also found in Cro-Magnon sites in France, Spain, and Italy. Ivory beads and animal tooth pendants were favored by

FIGURE 37.5 Venus figurine from the Paleolithic (This image comes from Wellcome Images, a website operated by Wellcome Trust, a global charitable foundation based in the United Kingdom).

the Cro-Magnon of France and Russia. Randall White is an archeologist who specializes in the study of Paleolithic art. He believes that the jewelry items of the Cro-Magnon artisans held a symbolic value to the people who made and used them. One reason for this conclusion is the complexity of the processes and the tremendous amount of time it took to make the items. Consider Aurignacion ivory beads for example. First, the ivory was shaped into a pencil-like rod several inches long. Next, the rod was divided into sections of equal length and grooved around the periphery. Then the pieces were snapped off to form individual beads. Then holes were drilled through each bead. Finally, abrasives were used to shape and polish each bead. It is estimated that each bead required three hours of labor to make and yet thousands of these beads have been found in Aurignacion sites.

37.11 MUSICAL INSTRUMENTS

Prehistoric bone flutes date back to at least 35,000 years ago and represent the first musical instruments ever discovered. The Divje Babe flute, carved from a bear's femur, may be the world's oldest instrument at 43,000 years old. It was found in a cave in Slovenia. There is a dispute over whether the flute was made by Neanderthals or Aurignacians. Another bone flute was found in Germany and is thought to be Aurignacian. It is a five-holed flute with a V-shaped mouthpiece and is made from a vulture's wing bone. It dates to 35,000 years ago. Probably the first musical instrument was the human singing voice.

37.12   COMMUNICATIVE SKILLS AND INTELLECTUAL LEVEL

What was verbal communication between Cro-Magnon people like? Obviously, there is no way to know the answer to that question for sure. However, we can deduce a lot about it from their behavior and interests. The Cro-Magnon cultures are identified by their tool and weapon artifacts, which seemed to have changed over time. They were obviously adapting rapidly to changes in available game and materials for fabricating tools and weapons. That ability to visualize the solution to problems and then fabricate a new tool or execute a new process requires superior communicative abilities. As communicative skills became more sophisticated, new problems became easier and faster to solve.

Spiritually, the Cro-Magnon were not only discussing deep topics like life, death, good and evil, but practicing symbolic rituals. The cave art, which persisted for 25,000 years or more, is one indication of it. The sculptures, figurines, musical instruments, and thousands of beads adorning burials are another indication of it. For example, at an open-air site in Russia dated at 29,000 years ago, 3000 ivory beads were found at the burial site of an adult and 10,000 ivory beads were found at the burial site of two children. They had been arranged head-to-head in their grave (Klein and Edgar, pp. 265–266).

37.13   SUMMARY

Human evolution cannot always be detected by just comparing fossil bones. Human intellect can change while the body stays the same. The Great Leap Forward is an example of that. The human hunter-gatherer culture had stayed the same for hundreds of thousands of years as judged by stone tools and other artifacts. Then a cultural change based on innovative adaptation began about 50,000 years ago, which matched the timing of the out-of-Africa migration. These ancestors of ours, previously adapted to a milder climate, not only survived in the colder, higher latitudes of Eurasia, but thrived and grew in number. Their hunting ability improved to the point where they diminished the number of their prey and it became harder to find. They used their ingenuity to find food elsewhere. They invented new weapons, traps, and nets to capture fish, seals, birds, and rabbits. Symbolism became important in their daily lives. Art renderings of animals, figurines, sculptures, jewelry, bone flutes, and decorated implements became commonplace. They grew in spiritual and aesthetic awareness. Humanity continued the innovative trend right up to today.

CHAPTER 38

AGRICULTURE AND CIVILIZATION

CONTENTS

38.1   SCOPE

Agriculture and domestication of animals began to spread worldwide beginning at about 10,000 years ago. Until then, humans lived as hunter-gatherers throughout the world. The new life style had its pluses and minuses as we shall see in this chapter. On the plus side, agriculture solved the problem of not having enough food. In fact, it even facilitated a population increase. On the minus side, it introduced an unhealthy life style where diseases and malnutrition weakened the participants. However, there was no going back once farming and ranching took over. Humanity was changed socially and genetically. Agriculture played a pivotal role in the evolution of human beings.

38.2   THE HUNTER-GATHERER AGE

The Homo lineage existed as hunter-gatherer societies for at least 1.5 million years and probably much longer than that. The ancient ones, whom we call *Homo erectus*, were the trail blazers for this life style. They adapted to life on the open ground, gave up their dependence on sanctuary in the tree, mastered the use of fire, fabricated and used sharp stone tools and weapons, and became coordinated in their defense against predators. Eating meat as part of their diet provided the high-energy food needed for this new life style. Their bodies had also evolved for endurance-type running by developing long legs, shorter arms, losing most of their body hair, developing abundant and efficient eccrine sweat glands, and evolving brain-cooling blood-circulation features. These attributes helped make them extraordinary runners. In fact, they could run prey down until it collapsed from overheating and exhaustion while not becoming overheated from the effort themselves. Even today, humans are extraordinary long-distance runners.

Most noticeable of the changes during the *Homo erectus* reign was a phenomenal increase in brain size. The australopithecines had existed for over two million years, yet their brain size stayed very similar to that of chimpanzees. In contrast, the brain size of the Homo lineage tripled from that of an ape. Superior intelligence was the most important survival tool of the evolving hunter-gatherer. How are we related to *Homo erectus*? The species *Homo heidelbergensis* was one offshoot of *Homo erectus*. This species is believed to the common ancestor to both *Homo neanderthalensis* and *Homo sapiens*. There may have been other human species yet to be discovered, but we think that all of them made hunter-gathering their way of life. As we have seen, these migratory hunter-gatherers, from as early as 1.8 million years ago, left Africa and occupied regions of Europe and Asia. In the course of their migrations, these versatile people adapted to many different climates. They even thrived in the harsh conditions of the Ice Ages in northern regions.

Hunter-gatherers lived in small groups, led a nomadic life, ate a balanced diet, were relatively disease-free, and had individual rights due to their status as equal contributors. Genetically, they were hard-wired to live by their wits, function as a team, and adapt to new environments. About

50,000 years ago, our species displayed a marked increase in using innovation to not only adapt to new environments but to thrive in them. We call this surge in innovative spirit, the Great Leap Forward. That same spirit led to the development of agriculture and domestication of animals. This change was so powerfully important that we call it the Agricultural Revolution.

38.3   THE FERTILE CRESCENT

The hunter-gatherer life style was freer and easier than that of a farmer or herdsman. So, why did these new ways of living ever arise? They probably arose out of necessity. It is speculated that edible wild plants were being over-harvested and that wild game was being over-hunted. The wild gazelle, especially, was diminishing in numbers, and it had been the hunter-gatherer's principal source of meat. Agriculture, which appeared around 12,000 years ago, changed that free and easy life style entirely. Although agriculture independently arose at various places in the world, it was the Fertile Crescent where it made its biggest impact (Figure 38.1).

The location of the Fertile Crescent begins at the eastern end of the Mediterranean Sea (The Levant) and continues eastward to the Persian Gulf. This crescent-shaped zone includes three great rivers. They are the Tigris and Euphrates rivers and this area was once known as Mesopotamia. The crescent also includes much of the Nile river and its delta. This area is Egypt. The Fertile Crescent is bounded by mountains on the north and by desert on the south. It is a moist and fertile zone between these hostile barriers.

38.3.1   THE IMPETUS FOR FARMING

Domestication of animals began in the Fertile Crescent with sheep, goats, cattle, and pigs. As for the domestication of plants, it developed as follows: The residents had already been eating and collecting a variety of seeds, but by 8000 B.C., einkorn wheat, barley, and chickpeas were domesticated as farming plants. Domesticating the wild cereals, wheat, and barley may have been a lucky incident. These cereals were already

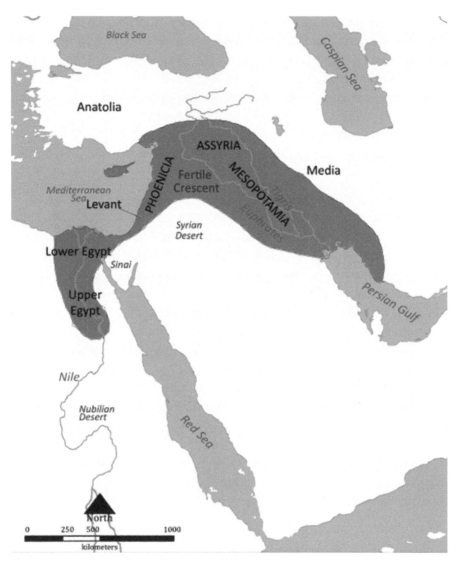

FIGURE 38.1   Map of the Fertile Crescent (Courtesy of Wikimedia Commons by Colt .55 [GFDL (http://www.gnu.org/copyleft/fdl.html), CC-BY-SA-3.0 (http://creativecommons. org/licenses/by-sa/3.0/) or FAL], via Wikimedia Commons).

growing there naturally, had the convenience of a short growing season, and produced huge yields. So mankind's first attempt at domesticating plants paid immediate rewards.

Once farming developed, there was excess grain to store, and people began to rely more and more on these reserves for something to eat during lean periods. Domesticated animals meant food would be available when hunting was unproductive. Cattle, sheep, and goats are a source for milk. Slaughtering animals for meat could be delayed until the herds built up sufficiently, but milk was always available. Finally, there was the pressing need to obtain more food to feed a growing population. Farming met that need but allowed an even larger population to exist than before. It was a process that fed on itself with the result of rapidly expanding populations wherever agriculture got established.

38.3.2 PASTORALISM

Many authors lump animal domestication, a.k.a. pastoralism, as a subset under agriculture, but I want to point out that although pastoralism and farming were concurrent events, they were often very different from each other. Crop farmers are limited to areas having moist fertile soil whereas pastoralists are able to graze their herds on a wider variety of terrain, including rocky hills and semi-arid deserts. Grazing animals such as sheep and cattle require large tracts of land to sustain them. In fact, they need to be frequently moved to fresh pastures. One of the common themes in western movies is hostile conflicts over rights to grazing land. That history of hostility may have originated in the sheepherders and cattle ranchers of the Fertile Crescent. These nomadic people were more inclined to using force than were the farmers or village dwellers.

38.3.3 THE NEOLITHIC REVOLUTION

Archeologists have documented the fact that the early farmers were solving the problems associated with harvesting and storing grain based on their recovery of ancient implements designed for these tasks. For example, sickles have been recovered with flint blades cemented into wooden or bone handles. Woven baskets were developed to carry the grain. Mortar and pestles and grinding slabs were used to remove husks, and plastered underground pits were fabricated to store the grain in a dry condition. Some of these relics date back to as early as 11,000 B.C. This transition from arrowheads for hunting to scythes and hoes for farming represents a major change in stone toolmaking; it is called the Neolithic revolution.

38.4   THE SPREAD OF AGRICULTURE

Jared Diamond, in his Pulitzer-Prize-winning book *Guns, Germs, and Steel*, points out that the east-west major axis of Eurasia gave it a significant advantage over Africa and the American continents, where each have a major axis that lies north-south. He is talking about the spread of agriculture from its point of origin. This simple concept is very important to understanding the history of the world during this transformational time. Crops can be successfully transplanted at the same latitude where growing seasons are similar but are more likely to fail when transplanted along a longitudinal line where growing seasons are quite different.

38.4.1   SPREAD FROM THE FERTILE CRESCENT

The spread of food production was faster in the east and west directions than it was in the north or south directions. The reason for this is that the climate is very similar at the same latitude. The change in seasons and hours of sunlight are similar, and the plants have adapted to these factors. However, higher or lower latitudes upset the plants schedule and may cause the crops to fail. The spread of agriculture from the Fertile Crescent was west into Europe, south into Egypt and Ethiopia, and east into Central Asia and the Indus Valley (Fertile zone within Pakistan and Western India).

Jared Diamond illustrated the spread of agriculture in radial waves emanating from the Fertile Crescent after 7000 BC. The wave reached Greece, Cyprus, and the Indian subcontinent by 6500 BC, Egypt by 6000 BC, central Europe by 5400 BC, Spain by 5200 BC, and Britain by 3500 BC. By 1 BC, Fertile Crescent crops had spread across the Eurasian continents. The zone reached from Ireland to Japan.

38.4.2 SPREAD FROM OTHER CENTERS

38.4.2.1 Africa

The spreading wave reached Ethiopia at an uncertain date, but independent crop domestication was underway there. The Sahel is the east-west lying zone in North Africa, which lies south of the Sahara Desert and is bound on the west by the Atlantic Ocean and on the East by the Red Sea. The Sahel is mostly grassland and savannah, with areas of woodland and shrub. The climate is hot, sunny and dry. Domesticated plants were developed there about 5000 BC. They were sorghum and African rice. Livestock is tended in the Sahel by semi-nomadic pastoralists.

The spread of food production in Africa went southeasterly from the Sahel and West Africa towards South Africa and East Africa.

38.4.2.2 Americas

Mesoamerica includes Mexico and down through Central America. Agriculture, mainly corn and beans, began in Mesoamerica before 3500 BC and slowly spread northward into North America. Cotton was also grown for its fiber there. Mesoamerica had no domestic animals other than dogs. Llamas were domesticated in Peru, but never made it northwards to Mesoamerica. Yet somehow manioc (a.k.a. cassava), sweet potatoes, and peanuts spread from South America to Mesoamerica. Some crops were independently developed in both South America and Mesoamerica (e.g., lima beans, common beans, and chili peppers).

38.4.2.3 Pacific Islands

Food production spread very rapidly from Eastern China into Southeast Asia, which includes: The Philippines, Indonesia, Korea, and Japan. The package consisted of Fertile Crescent crops plus additions such as bananas, taro, and yams. Domestic animals included chickens, pigs, and dogs. This, being an east-west spread, penetrated 5000 miles in only 1600 years, reaching all the way to Polynesia.

38.4.3   OTHER INNOVATIONS IN THE FERTILE CRESCENT

In addition to the development of cereal crops and domesticated animals, other innovations in the Fertile Crescent soon followed. The wheel was invented and first used in ox-drawn carts for moving grain. Writing was invented for bookkeeping purposes related to farming, but soon found many other uses. Fruit trees were domesticated and alcoholic beverages (i.e., beer and wine) developed as an industry.

38.5   HOW AGRICULTURE GENETICALLY CHANGED HUMANS

Gregory Cochran and Henry Harpending did an in-depth analysis of how the change in life style of going from hunter-gatherer to farmer produced severe stresses on the farmers. The stressors became less severe over many generations via genetic change and altered traits. Let us consider them next.

38.5.1   INFECTIOUS DISEASES

Farming resulted in larger groups of people living closer together in a permanent location, and often living quite close to their domesticated animals. Waste from humans and animals often contaminated drinking water, and diseases from humans and animals spread through the community with often fatal consequences. Grain and food waste attracted rats and mice, which occasionally carried typhus and bubonic plague. Due to these stressors, genes favoring a stronger immune system increased in frequency due to natural selection.

The greater size and density of human populations are often correlated with the occurrence of contagious diseases. These factors are important in determining if a disease continues to infect some part of a large population and keep the disease continuously alive in that location. However, the origin of disease is mostly associated with living in close contact with farm animals. We know this because such animal contact was common in the Fertile Crescent where numerous diseases originated. Yet few to no diseases originated in the American cities. They had essentially no ani-mals to domesticate. When Europeans invaded the Americas, contagious diseases such as smallpox decimated the native populations. Europeans, on the other hand, had no such problem succumbing to American diseases because they didn't exist.

38.5.2   NUTRITIONAL ADAPTATIONS

Hunter-gatherers ate a much healthier diet than people in a farming com-munity. Farming produced more food than before, but it was far less var-ied, much heavier in carbohydrates, diminished in proteins, and did not provide all of the nutrients required by the human body. The new diet keeps a larger population alive, but at a price. The average height of farm-ers dropped by five inches when compared to the hunter-gatherers of the time. Malnourished individuals were more susceptible to disease, and infant mortality rose as agriculture replaced the hunter-gatherer life style. The cereal-rich diet led to vitamin deficiency with diseases like beriberi, pellagra, rickets, and scurvy. These diseases were unknown to hunter-gatherers, who ate a varied diet. Table 38.1 lists the symptoms of these diseases and the missing vitamin or nutrient that led to the disease.

TABLE 38.1   Kinds and Effects of Malnutrition

Disease	Symptoms	Vitamin Needed and Food Source
Beriberi, wet and dry	Damage to heart and circula-tion (wet)	Thyamine: beans, vegeta-bles meat, and whole grains
	Damage to nerves and muscles (dry)	
Pellagra	Diarrhea, colitis, dermatitis, dementia	B-3 (niacin): proteins
Rickets	Skeletal disorder, seen most in young children, bones weak and soft	Vitamin D, calcium, and phosphorus: sunlight, eggs, fish, and milk
Scurvy	Anemia, debility, exhaustion, edema, and loss of teeth	Vitamin C: Fruits and vegetables

In the Americas, maize became one of the main food staples, but tooth decay and anemia soon followed. Maize is low in iron, which the body needs to make hemoglobin.

38.5.3   LACTOSE INTOLERANCE

Mammals, including man, have a genetic blueprint that produces an enzyme (i.e., lactase) for digesting the sugar, lactose, in their mother's milk. Lactase is produced during the breast-feeding years, but then it shuts down. In other words, adult mammals are lactose-intolerant; milk is no longer digestible. In fact, drinking milk could cause bloating, cramps, flatulence, diarrhea, nausea, or even vomiting. Yet the herders of milk-producing animals (e.g., cattle, goats, etc.) had a source of food that was readily available. The younger children could probably digest it, but it made adults ill. Diversity played its normal role in genetic change. Some individuals were more lactose-tolerant than others and this gave them a survival advantage. Milk contains vitamin D and that may have been a survival factor for the higher latitudes, where low sunlight diminishes production of vitamin D. Milk drinkers were healthier than their peers because they could digest a normally available food. The population gradually became more lactose-tolerant because the trait was passed to the milk drinker's children and so on through the generations. Surveys indicate that lactose-tolerance today correlates well with regions of the world where animal milk was harvested. Over 70% of northern Europeans are lactose-tolerant. The mutation producing lactose tolerance has been traced back to its origin 8000 years ago. There is a very high percentage of lactose-tolerant Europeans, whereas it is rare in other parts of the world. Milk-drinkers had a competitive advantage over others; for one thing, they were taller. Many pastoralists learned to specialize in dairy cattle, as it was far more efficient than raising cattle for meat alone. I remind myself as I write this paragraph that I grew up in the dairy country of northern Illinois. Classmates of mine from school lived and worked on dairy farms. Wisconsin, the state just north of us, was even more dedicated to dairy farming. Milk is a vital part of our lives as it was for the northern European farmers who brought the tradition to America.

38.6   HUMAN EVOLUTION DURING THE AGRICULTURAL REVOLUTION

The invention of agriculture had powerful evolutionary consequences for mankind. On the one hand, food was plentiful and the threat of starvation had diminished. Population size grew as a result. On the other hand, malnutrition and contagious diseases were a new threat to those living as farmer or villager. These threats were having an accelerated selection effect on humanity because they had such deadly consequences. Diversity is the key to survival of the group. Those having the best genes for dealing with these threats passed those superior genes on to the next generation. Those with the worst genes were eliminated. The first few generations of farmers, who came down with diseases like measles, mumps, chickenpox, etc., were as unprepared as the poor native Americans, who died in troves from these diseases many years later when Europeans set foot to American soil. Yet those of us descended from Europeans have immunity to these diseases or at least do not die from them. I, like many of you, had many of these childhood diseases when I was school age. I survived all of them.

38.7   PERSONALITY AND MENTAL CHANGES

Cochran and Harpending also propose that agriculture had the effect of radically changing the personalities of the farm workers at a genetic level over the generations. Just as domesticated animals have become more docile than their wild counterparts, so farmers became domesticated by the elites who gained power over them during this time. The idea of chiefs or bosses governing their behavior was unacceptable to hunter-gatherers and still is today. The rise of the elite developed as a consequence of agriculture. The farm has value to its owner and he is unwilling to walk away from it. Yet, he is vulnerable to those who would take his possessions forcefully from him. The farmer became dependent on the elite for protection, but he got taxed for the maintenance of it and lost the personal freedom he once enjoyed.

Farmers have to be selfish and willing to suffer in the short run in order to benefit in the long run. For example, they must store seed grain in order to have seeds for next year's crop. They dare not consume it even if their family goes hungry. Hunter-gatherers never thought this way. If they killed a large animal, everyone joined in the feast. The meat would go bad if it was not eaten, so why hoard it. Humans, engaged in farming and pastoralism, were evolving different personalities than they had as hunter-gatherers. Those who could submit to powerful masters sacrificed in the short run to benefit in the long run, and resisted the urge to share their gains immediately with others survived, even thrived as farmers. Those who could not so adapt were at a disadvantage and passed fewer genes forward.

38.8 SUMMARY

An agricultural life style gradually displaced the hunter-gatherer life style around the globe. There still are a few hunter-gatherer societies left, but the contrast between them and the modern city dweller is immense. The benefits of the agricultural revolution were so important that there was no going back to simpler times. Besides, the booming population of humans and diminishing population of game animals precluded that possibility. Eurasian agriculture began in the Fertile Crescent about 10,000 B.C. It was able to spread in an east-west direction across Eurasia because imported plants take better if the latitude is the same. Agriculture spread more slowly in Africa and the Americas because the major axes of these areas are north-south.

The adaptation to denser populations, no longer migratory, living close to animals, eating a nutrientally inadequate diet, was brutally difficult. Diseases jumping from animals to humans was one problem. Living in one location where waste products and vermin thrived was another. Going from the healthy hunter-gatherer diet of meat, berries, fruit, and grains to a mainly grain diet generated diseases caused by vitamin and mineral deficiencies. It took many generations for humans to genetically adapt to the stresses of disease and malnutrition. Even our personalities had to change to adapt to a hierarchal society. The hunter-gatherers had greater independence than the farmers and village dwellers. They also readily shared whatever they had. Farmers had to be hoarders, who selfishly guarded their crop seed in order to continue producing crops year after year.

CHAPTER 39

THE JOURNEY FROM APE TO MAN

CONTENTS

39.1 SCOPE

In the previous chapters, we learned that despite the great differences between humans and chimps, they are our closest relative. DNA evidence tells us that we had a common ancestor with them 5–7 million years ago. So how did we become so very different from them? We have used fossil and relic evidence to describe our hominin ancestors and surmise how they lived. We have used evolutionary theory and principles to attempt to understand the humanization over time. Finally, we have employed DNA science to track our migrations and interbreeding history. In this chapter, we pool that knowledge and provide an answer to that original question of how we could have evolved from apes.

Apes proliferated during the Miocene epoch (23.0 to 5.3 mya), but then climate change drove our ape ancestors out of their comfortable tree-dwelling life style and further and further from it over millions of years. The australopiths were apes, who physically evolved as they mastered the art of biped walking. The Homo lineage evolved from one of those species of australopiths some 2.5 million years ago. We are the last surviving species in the Homo lineage but our numbers exceed seven billion individuals. Let us revisit the transformations that step-by-step made us less a chimp and more a human. Finally, we will assess the state of humanity now and where it might be headed.

39.2   HOW WE ARE DIFFERENT FROM CHIMPS?

When I started writing this book, I was fascinated with the glaring conflict between DNA evidence that proved we are more closely related to chimps than any other living animal and the huge physical and mental differences between chimps and humans. Think of the paradox! DNA from two related species is very nearly identical, and yet one genome yields a fur-covered, small-brained, tree-adapted, quadrupedal ape with an inability to speak, write, or think creatively, whereas the other genome yields a hairless, deftly bipedal, large-brained human being, who can walk, talk, write, and invent spaceships that land on and return from the moon.

The bonobo is another near relative to the chimp; a sister species in fact. After a million years of separation, the two species are still physically hard to tell apart. We might also note that gorillas, which are more distantly related to chimps than bonobos, are also fur-covered, small-brained, quadrupedal apes with an inability to speak, write, or think creatively. We are closely related to them too. So, why are humans so different from these African apes and what evolutionary factors made our amazing transition occur?

39.2.1   ADAPTATION TO A CHANGING ENVIRONMENT AS THE EARTH GREW COLDER

The common ancestor of chimps and humans lived some 5–7 million years ago and was probably very similar to today's chimps. During the Miocene (23 to 5.3 mya), the world's climate had been very favorable to the proliferation of apes. As many as 30 species of apes existed and they inhabited large regions in Africa, Asia, and Europe. The Miocene epoch had produced extensive forests, rich with foods that apes like to eat, like fruits and nuts. However, environmental conditions got worse for the apes, and their numbers and range began a long decline. What followed was millions of years of progressively colder and drier conditions. Figure 39.1 shows the temperature changes and accumulation of ice during the Cenozoic era. Notice the marked cooling that occurred during the last seven million years and the accumulation of ice.

The humidity of the planet dropped as more and more water became locked up as glacial ice. These climatic changes affected plant life around the globe. The tropical forests were in decline, and some apes adapted to the more open woodlands and grasslands by becoming proficient bipedal walkers and learning to find new kinds of edible plants. Survival now meant spending more time on the ground and less in trees.

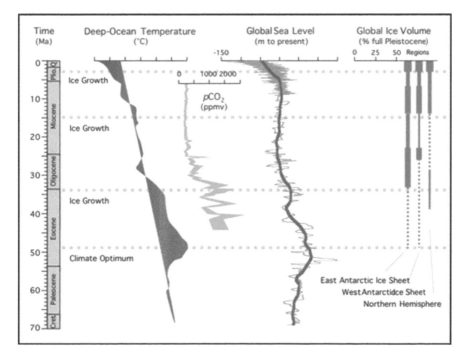

FIGURE 39.1 Climate change in the Cenozoic era (Courtesy of Wikimedia Commons, By Diekmann, B./AWI [CC BY 3.0 (http://creativecommons.org/licenses/by/3.0)], via Wikimedia Commons).

39.2.2   EVOLVING INTO A MORE VERSATILE APE

Those apes who broke with their former dependence on life in the trees became the only mammal capable of habitual locomotion in an upright position. These bipedal apes were adapting to a changing environment and their new capabilities gave them advantages. The benefits of this adaptation were twofold: (i) They were more versatile than their tree-bound cousins in being able to survive in different kinds of terrain, and (ii) they were better positioned than their tree-dwelling cousins to making further adaptations to the continually changing environment. Their progeny, having upright walking ability, would be better adapted than the tree-dwellers to survive in an environment becoming very different from a tropical forest. Their long arms made it still possible to quickly climb trees when seeking safety from predators, but they were also evolving into proficient walkers and ground-dwellers. These evolutionary changes involved extensive skeletal modification including: an upright posture, a forward-looking and centrally balanced skull, a much-modified pelvis, angled legs, and feet with inline toes. Collectively, we call these bipedal ape ancestors, the Australopiths. They persisted for millions of years in eastern and southern Africa.

39.2.3   ROBUST AUSTRALOPITHICINES, A.K.A. PARANTHROPUS

Meanwhile the chimps, living in those few remaining forest clusters, were becoming even more specialized in their adaptation to a tree-bound existence. However, climate change continued to alter the environment of Africa, producing savannahs and deserts. Survival for the bipedal apes required that they undergo even greater evolutionary change. By 2.5 million years ago, the Australopiths found it increasingly difficult to find food due to the much drier and open environment. One group of them evolved in the direction of bigger molars, jaws, and anchoring points for chewing muscles. We deduce that they developed these giant molars and chewing muscles in order to process very tough, fibrous food. We call these particular man-apes, the robust Australopiths or some anthropologists assign them to a new genus *Paranthropus*. They did not evolve vastly larger brains as did their Homo lineage contemporaries. The *Paranthropus* persisted for about one million years but went extinct about one million years ago (Figure 39.2).

39.2.4   THE HOMO LINEAGE APPEARS IN THE FOSSIL RECORD

Concurrently, another line of Australopiths thrived by scavenging meat from dead or dying animals. They used sharp stones to sever meat from the carcasses. This genus, which we call "Homo," began to gradually resemble modern man more and more over time. The fossil record is incomplete during this transitional period of Australopiths evolving into Homo, and the topic of fossil assignments is controversial amongst paleoanthropologists. The species, *Homo habilis*, is sometimes mentioned as the first of the

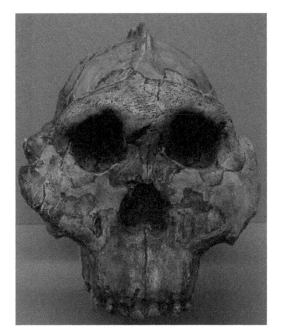

FIGURE 39.2   Skull of *Paranthropus bosei* (By Thomas Roche [CC BY-SA 2.0 (http:// creativecommons.org/licenses/by-sa/2.0)], via Wikimedia Commons).

Homo lineage, but the fossil evidence is sketchy. The stone tool evidence is convincing though. The earliest stone tool culture is called "Oldowan," and the implements are quite crude.

Meat in the diet had a transformational effect on the Homo lineage. The brain size increase was the most remarkable of the changes. It doubled and nearly tripled in size over the next 1 to 2 million years. Legs became longer, arms became shorter, the gut got smaller, and hairlessness developed. The Homo lineage became ever more like modern humans. Perhaps speech developed during this time, and these hominins became ever more efficient hunter gatherers. Inventions played a big role in making these adaptations possible. Stone tools and weapons became essential to their life style. The ability to cut chunks of meat from a carcass and escape with the meal to safety before predators returned was dependent on sharp tools. Taming of fire, cooking of meat and vegetables, and using fire to scare off predators at night freed our ancestors from the need to sleep in tree nests anymore. These latter adaptations were in the time of *Homo erectus* (sometimes called *Homo ergaster* in Africa).

39.2.5   *HOMO ERECTUS BECOMES A RUNNER*

At some point, our scavenging ancestors became hunters in their own right. Crude stone tools, dated at 2.5 million years ago, have been associated with *Homo habilis*. These cutting tools would have been essential for hacking off chunks of meat as fast as possible before predators returned to their kills. *Homo erectus* evolved from these early scavengers and as a result of a meat diet underwent significant physical and mental changes. His brain size increased, legs grew longer, gut grew smaller, but perhaps most significantly, he developed the ability to run large game down to their death. Mammal bodies heat up from prolonged running. Beyond a certain point, the mammal will collapse from exhaustion and overheating. The mammal brain is the weak point. It can only tolerate so high a temperature before dizziness and confusion occur. Human evolution during this period involved major changes to overcome these cooling problems: *Homo erectus* developed blood circulation routing that efficiently cooled the brain. He also developed a more efficient sweating method than used today by apes. It involved losing his fur coat and installing a system of numerous eccrine glands in the skin. His legs grew longer and his gait more efficient. Humans today are still excellent runners. Think of marathon competitors, for example.

Homo erectus goes back about 1.8 million years and his fossils have been found in Africa, Europe, and Asia, However, the stone tools he left behind are overwhelming evidence of his presence. After 1.7 million years, a new stone culture displaced the cruder Oldowan culture. It is called the Acheulean culture and it is far more advanced than the Oldowan culture it replaced. The hand axes seemed to have a special significance to these people because they have been found by the hundreds. The mental abilities of *Homo erectus* were obviously increasing even at this early date in his reign. The most noticeable physical and mental changes in our evolutionary journey from ape to human took place during this period.

39.2.6   HOMO ERECTUS ADAPTS HIS ENVIRONMENT TO SUIT HIM

Homo erectus had a bigger brain than the Australopiths before him, yet his brain size continued to grow even bigger throughout his over 1.5 million years on Earth. Bigger brains came with a high price. Childbirth was mortally dangerous and child-raising was more difficult because the infant was helpless for a longer span of time. Bigger brains are expensive to the individual too. They take more energy to operate. So there must have been large benefits for having big brains, considering its costs. Better stone toolmaking skill was one apparent benefit of a larger brain. Sharp cutting implements were vital to the successful scavenger, who made off with a severed hind quarter before the predator returned to its kill. Scrapers to prepare the pelts and axes to cut tree limbs were also important. Animal pelt clothing allowed them to live at different latitudes without freezing to death. Taming of fire was another innovation resulting from brainpower. Not only did campfires provide warmth at night and protection from predators, but fire led to cooking, which led to better nourishment. *Homo erectus* was capable of surviving on three different continents. His big brain made that remarkable feat possible.

39.2.7   THE HOMO LINEAGE LEARNS HOW TO COMMUNICATE

The world of ideas was expanded during the Homo lineage too. Planning and communicating were essential to the early hunter-gatherers on the savannahs. Sign language played an important part, but speech was a superior way to discuss different scenarios in order to select the best strategy and assign individual jobs to members of the hunting party. Language and mating were strongly linked as they are today. Human sexual selection and speech may have been the strongest influences driving brain growth as more intelligent males and females had more capable offspring, whereas the least intelligent suitors did not. Sexual selection played a large role in modifying both our physical features and our personalities. We have seen how physical attributes like prominent breasts, hidden ovulation, and boneless, flexible, large penises resulted from human sexual selection. The influence of female sexual selection acting on the males over many generations gradually changed the personalities of those men. The females were selecting for competent providers and protectors, yet also for kind, considerate, and humorous mates and caring fathers for their offspring. They needed to reject mean, selfish, and foul-tempered males. Over time, human personalities improved as sexual selection culled the worst of them. Of course, the influence of males selecting female mates had evolutionary effects on women as well. Males selected for fitness, fertility, child-raising skills, being affectionate, and meal-making ability. Those selections changed feminine traits too.

39.2.8   OCCUPYING THE EARTH

Africa is where all of the early hominin fossils have been found. The evidence is overwhelming that Africa is the birthplace of the human race. Our ancient ancestors eventually did venture out of Africa and occupy the rest of the habitable Earth. Human evolution had to advance to the point where hominins could survive on the ground without needing the safety of trees. The use of fire to keep predators at bay made this possible. The ability to hunt big game assured a source of nutritious food in unknown terrain. So far, fossil evidence of hominins outside Africa identifies them as of the

Homo lineage and dates back as far as 1.8 million years ago. *Homo erectus* was the first "out-of-Africa" pioneer and he probably made the exodus by simply following game. His fossils have been found as far east as Java and China and as far north as Dmanisi, Georgia.

Heidelberg Man, and the species he represents, evolved from *Homo erectus* about 800,000 years ago. This species, *Homo heidelbergensis*, is intermediate between *Homo erectus* and modern humans, but also intermediate between *Homo erectus* and the Neanderthals. His fossils have been found in Africa and in Europe. His brain size averaged around 1200 cc, showing that he was very close to modern humans in that department.

39.3   OUT OF AFRICA MIGRATION OF *HOMO SAPIENS*

Our species, *Homo sapiens*, has been in existence for at least 200,000 years based upon genetic and fossil evidence. A recent fossil find is said to put our reign back to 300,000 years. The use of mtDNA backtracking to identify mtEve and the use of y-chromosomal backtracking to identify yAdam showed that a wealth of information lies within the DNA of living people around the world. These same tools have proven useful for tracking the migrations of our species out of Africa some 50,000 years ago. New mutations in the migrating groups have provided us with markers, which are evident in the progeny of the regional population, but missing in the ancestors to those wanderers. Although *Homo erectus* got to Asia before us and Neanderthals got to Europe before us, we were the original founders of the America continents and the Pacific Islands. DNA evidence also tells us that our ancestors interbred with other human species such as Neanderthals and Denisovans. Nowadays, entire genomes are routinely sequenced for individuals from all over the globe. The story of human prehistory is becoming available due to DNA technology at an increasing clip.

39.4   BECOMING MORE INVENTIVE

One of the important lessons learned from studying the "Out-of-Africa" migrations is that these explorers were the most innovative humans ever seen before. The two species of humans, which we know the most about, are the Neanderthals and the modern humans. There were other Homo species around, like the mysterious Denisovan man, the Hobbits of Flores Island, and *Homo naledi* of South Africa, but we have just started learning about them. Now, the Neanderthals had evolved special physical adaptations for the cold climate of Europe and the Middle East over many thousands of years. *Homo sapiens*, in contrast, were adapted to an equatorial climate. These two species of the Homo lineage first encountered one another around 100,000 years ago in the Levant and then again between 40,000 and 30,000 years ago, when our *Homo sapien* ancestors first migrated into Europe. Fossil evidence tells us the Neanderthals hunted using an ambush style attack on big game. Their bones showed the kind of damage expected from encounters with battling animals. Our ancestors' bones did not exhibit the same damage, nor were they as sturdily built as the Neanderthals. Our ancestors' hunting strategy involved hurling spears from a safer distance. Their invention of the spear thrower made this hunting style even more effective.

About the time that the Neanderthals finally went extinct (about 30,000–40,000 years ago), our ancestors were already demonstrating a sharp increase in inventiveness. There may have been a genetic reason for this marked shift in behavior. We carry a small percentage of Neanderthal DNA in our genome, which came from interbreeding. Perhaps some of those Neanderthal alleles were beneficial to us and caused secondary changes in brain functions. Whatever the cause, the Cro-Magnon people of this time began making jewelry, figurines, new tools and weapons using new materials, creating cave art, and other creations. This burst of creativeness continued from this time forward. Over the next 20,000 or so years, they invented the atlatl, the bow and arrow, fish nets and barbed spears for fish, pottery, agriculture, domestication of animals, and horseback riding. Moreover, symbolism became highly important to them as demonstrated by ornate burials, statuettes, decoration of tools and weapons, jewelry, etc. Symbolism enhances creative thinking and may have provided the impetus behind the Great Leap Forward.

Our ancestors became far more innovative than they had ever been before. This began during their migration out of Africa. This Great Leap Forward was a turning point for humanity. It led to large population increase and the emergence of a new type of human. We were no longer just a creature that survived, but one with a different value system. There was a spiritual appreciation for life expressed in art, jewelry, sculptures, music, burials, and rituals. They started a culture that became integral with what it means to be human. We have the benefits of it today.

39.5   AGRICULTURE

The Homo lineage existed as hunter-gatherer societies for well over one million years, and there are still people living that way today. Hunter-gatherers lived in small groups, lived a nomadic life, ate a balanced diet, were relatively disease-free, and had individual rights due to being equal contributors. Agriculture, which first appeared 10,000 years ago, changed that scenario entirely. On the plus side, agriculture and domesticated animal farming increased the food supply markedly. This allowed for the rise in populations, permanent residency, and greater division of labor. On the minus side, the new diet of mainly grains was unhealthful, and the larger, denser populations led to pestilence and new diseases. Some diseases were spread due to ignorance of germs and the benefits of sanitation. In other cases, farmers lived closely to their animals and diseases crossed from animal to humans. Farming introduced new stresses to its human participants: stresses of pestilence, germs and poor diet. Those with weak immune systems died out, whereas those with strong immune systems prevailed. The short-term effect of farming was severe. It took many generations for the farmers to develop immunity to the diseases and develop more healthful practices.

Meanwhile, the pastoralists thrived and prospered. These people were herders, who continually drove their herds of sheep, goats or cattle to new fresh pastures. They ate meat, which solved many of the nutritional problems, and they did not live in their own filth as did farmers. These herders became healthier due to their new life style. These pastoralists evolved the ability to digest milk as adults. Milk was available from the cattle, sheep, and goats that they herded, and milk had a dramatic beneficial effect for those able to digest it. A favorable mutation made this possible. The favorable gene spread through the regional population quite quickly.

39.6   THE RISE OF CIVILIZATION

The Fertile Crescent was the place where agriculture began and spread. It is also known as the cradle of civilization. The rise of villages and even cities was made possible by agriculture. The reliance on innovation that marked the Great Leap Forward was even more important during this phase of our past. Stone scythes were developed to more efficiently harvest wheat and other crops. Baskets were invented to carry the grain. Pits with plastered walls were invented to store the grain in a dry state. The wheel was invented to move grain and other products more easily. Oxen-drawn carts made transport less of a chore. Writing was invented to record transactions of commerce, but quickly evolved into a new method of communication in general.

History books pick up the story at this point, and names like Samaria, Mesopotamia, Babylon, and others become familiar. Suddenly, the human evolutionary story becomes too expansive to fit in this book.

39.7   SUMMARY

The first impetus to change for the tree-dwelling apes was global climate change and the disappearance of much of the tropical forest supportive to an ideal ape environment. Bipedal apes evolved during this period (4–5 mya). These Australopiths evolved an upright stance and underwent skeletal changes to become better and better bipedal walkers. The cooling continued with the emergence of extensive grasslands and deserts. Two new genera of hominin replaced the Australopiths about 2.5 mya, *Paranthropus* and *Homo*. The first group specialized in fibrous foods, but the second group focused on meat. This group evolved into big-brained, tool-using hunter-gatherers, whereas the first group eventually went extinct. Our species, *Homo sapiens*, evolved from this lineage some 200,000 years ago. About 50,000 years ago, our ancestors left the evidence that they were the most innovative people ever seen. They used intelligence to solve survival problems and thrived as a result. They migrated out of Africa and filled all the landmasses of the Earth, save Antarctica. Around 10,000 years ago, they began the agricultural revolution, which led in turn to cities and civilization.

CHAPTER 40

WHAT THE FUTURE HOLDS

CONTENTS

40.1 SCOPE

We now understand how evolution works to modify a species when its environment changes or other new threats to its survival appear. Humans today represent a long series of bodily modifications over millions of years, starting from a tree-dwelling ape. We, alone among the animals, can control our destiny using innovation to solve new problems. Medical science has removed most of the worries about injury and disease. Many of us would be dead today were it not for medical science. So is evolutionary change no longer applicable to our species? Examining this question is the central theme of this final chapter.

40.2 WILL HUMANS CHANGE IN THE FUTURE?

40.2.1 HOW CHANGE OCCURS

Some anthropologists declare that human evolution has ceased and we will stay more or less the same in the future, barring a catastrophic event.

For example, evolutionary biologist Stephan Jay Gould wrote that there has been no change in humans in the last 40,000 or 50,000 years. Gregory Cochran and Henry Harpending see it very differently. They believe the pace of human evolution is at an all-time high and will go even faster if our explosive population growth continues. Their argument goes something like this: In order for a new allele to become part of the typical human genome, it must be formed by a mutation in the germ line, it must be beneficial, and it must noticeably increase its frequency in future generations. Mutations can be harmful, neutral, or beneficial. Harmful mutations get removed from the genome by natural selection. Neutral mutations have nothing to spur their increase in frequency, but beneficial mutations do. Natural selection works on populations to optimize their fitness relative to what they eat and where they live. Perhaps Gould was wrong about human evolution creasing 50,000 years ago. Let us consider the case of skin color to illustrate recent human evolutionary change.

40.2.2   WHY DO HUMANS HAVE SUCH DIFFERENT SKIN COLORS?

The *Homo sapiens*, which left Africa during the Great Leap Forward, must have been dark-skinned and similar in appearance to the indigenous people who live in Africa today. Remember that they were the ones to settle in Europe, India, Asia, and the American continents, yet think about how different each of these people are today from one another. Europeans tend to be swarthy colored in the Mediterranean countries, but fair-skinned in the northern countries. Northern Asians are fair-skinned too, but due to a different mutation than found for Europeans. Southern Asians, East Indians, and Pacific Island people tend to be darker skinned than the northern Asians. American natives also exhibit a range of colors depending on the latitude. So how did those changes occur and what evolutionary factors were involved in causing it to happen?

The answers to these questions are beautifully told in Nina Jablonski's book, *Skin: A Natural History.* She clearly explains what is at stake for us bare-skinned humans in regulating the UV radiation that our body receives from the Sun. On the one hand, we must protect ourselves against excessive UV exposure, but on the other hand, we must have enough UV radiation to allow the body to synthesize vitamin D in our skin. What is at stake is our ability to successfully reproduce ourselves, because the incorrect level of UV radiation will make successful reproduction impossible.

One of the ways that reproduction is compromised is by destruction of folate, a chemical vital to the production of DNA. Longwave UV radiation (UVA) destroys folate, and without folate, the DNA production vital to growing a fetus is not possible. However, the body has a solution to excessive UVA, namely: melanin. This biochemical can block the UVA from penetrating into the skin with the result that folate is not destroyed. The UVA problem is most severe at the equator and diminishes as one moves closer to either the North Pole or the South Pole. So it is not surprising that the darkest-skinned people (i.e., those with the most melanin) live near the equator and that skin color becomes progressively lighter as one moves closer to a pole.

The opposite problem to excessive UV exposure is too little UV exposure. Our dark-skinned ancestors from Africa faced this very problem as they migrated to the higher latitudes. They were unable to manufacture enough vitamin D and that problem also interfered with reproduction. Insufficient vitamin D impairs calcium metabolism, which means babies are born with soft bones unable to support weight. The disease is called rickets.

So we have a Goldilocks scenario in the case of blocking UV radiation. Too little blocking and we cannot reproduce our own kind, and too much blocking and the same problem exists. There is a range of UV blocking that is just right. However, UV radiation varies from most severe at the equator to least severe at the poles with intermediate severity in between. Melanin content in the skin establishes itself in stable regional people at the optimum level over many generations through natural selection.

40.2.3 MUTATIONS RATE VERSUS POPULATION SIZE

Mutations occur due to an error in gamete production by the body or by cosmic rays, and they are rare events. However, the larger the population, the greater the likelihood of a mutation occurring in a given period of time. Table 40.1 shows the world's population at various times in the

TABLE 40.1   Human Population Growth Trends

Year	60,000 BC	3000 BC	1927 AD	1974 AD	1999 AD	2016 AD
Population	0.25 million	60 million	2 billion	4 billion	6 billion	7.4 billion

past. The first two entries are significant in pointing out the influence of population size on mutation rate. Back in the days when the population was about one quarter million people, it took about 100,000 years for a favorable mutation to establish itself. However, by 3000 BC with the population at 60 million, it only took 400 years (see, *The 10,000-year Explosion*, page 65). I would guess that they are happening on a daily basis or faster with today's population size. The time for an advantageous allele to spread through the population is also hastened by the population size. Its frequency increases exponentially with time in a well-mixed population. In other words, it would only take twice as long for the favorable allele to spread through a population of 100 million as it did to spread through a population of 10,000 people. Thus, with mutations occurring at a rapid rate due to our huge population and an exponential rate of spreading, we are evolving faster as a species than at any time in our evolutionary history.

40.2.4   NATURAL SELECTION IN MODERN MAN

Today, the stressors affecting us are many of the same bad actors as in the past, namely, starvation, toxins, germs, viruses, and parasites. These are the agents that cull the weak and advance the strong. Avoiding predators is no longer important to human survival because we have eliminated them or reduced their populations significantly. Somehow predators have developed a healthy fear of man. It seems to be hard-wired into them to intentionally avoid us. At least this is true near civilized areas. Starvation is rarely a problem in the advanced countries of the world but still exists in the poorer countries, especially those undergoing civil strife. Malaria and other parasitic diseases have been eased but not conquered in many regions of the world. At least, we now know what causes it and what precautions can be employed to reduce infection. Toxins are probably a bigger problem than they were at an earlier time. We now have cities with millions of inhabitants living near industrial plants, which generate toxins in the normal course of business. I have seen articles showing a map of the areas of the USA where chemical processing is densest. Next, they showed a map where cancer cases are most prevalent. The two maps are superimposable. Finally, bacterial and viral diseases can be fought by modern medicines, but the Red Queen effect is always at work undoing our progress in defeating these diseases. Moreover, new versions of the influenza virus hit us every year and sometimes kill millions as they spread. So the bottom line is that although progress has lessened the dangers around us, they never entirely disappear. Another takeaway is that the poorest humans suffer the most from all of these stressors, and consequently they are evolving more rapidly than the rest of the population.

It is in the mental realm that human evolution seems to be making the biggest adjustments. Intelligence is of great importance for humans. A slight difference in intelligence can make a large difference in being able to successfully raise a family in an ever-increasingly complex world. In today's world, it pays to be likeable, fast to learn new things, and be skilled in selling your ideas. People with those skills can earn a better living and provide for their children better. On the other extreme, people lacking those skills will earn less, be unemployed more often, and will not have the resources to provide for their dependents.

One thousand years ago, life expectancy was about 40 years. Technology has changed that drastically, and as a result changed human society. Humans live into their seventies on average these days. One effect of greater longevity is that grandparents are an increasingly important factor in the human story. Grandparents not only raise their children to adulthood but continue the support to their children's children. Are genes developing to make us want to be good grandparents?

40.2.5   GENETIC CHANGE IN RECENT TIMES

Favorable mutations are happening in the world population at a record rate today simply because there are so many people. It has been estimated that our evolutionary pace has increased 100 times in the last 10,000 years. It is difficult to predict how this is changing and will change future human beings. It still takes multiple generations for these favored genes to rise in frequency in the population. In other words, it will take generations for us to see new traits in the majority of people.

However, we are able to compare the important genes over time. Those scientists who study genetic change report that recent evolutionary change in humans is associated with smell, reproduction, brain development, skin color, and immunity to disease and parasites. For example, the human brain has been getting smaller for thousands of years. Brain size used to average around 1500 cc and now it averages 1350 cc. One explanation is that the brain is getting more efficient and doesn't need to be as big as it once was. Genetic changes that arose thousands of years ago have not stopped their effects in causing new or enhanced traits to appear. Light-colored skin, hair, and blue eyes are relatively recent changes that are apparent. Scientists have shown that the evolution to lighter skin happened separately in Asia and in Europe. Skeletal changes have been occurring too. There is a continuing trend towards a lighter frame. Consider brow ridges, for example. The Australian aborigines still have them as well as a thicker skull than other humans. European skulls from 3000 BC sometimes have brow ridges as well. Jaws are getting smaller faster than the teeth they contain. I have had two of my wisdom teeth extracted because they were impacted. There was inadequate room for them and they pushed against the adjacent molars at an angle. I am typical in having problem wisdom teeth. Eventually, this third molar will be eliminated entirely due to evolution.

The human adaptations to the agricultural age are still working their way through the world population. These include acquiring immunity to communal diseases, developing tolerance of lactose in milk products, adapting to high carbohydrate diets, etc. Not only physical changes occurred during the agricultural revolution. Personality changes occurred as well as humans made the transition from an equalitarian society to a stratified one. Personality changes affected are selfishness, forsaking immediate rewards for future rewards, controlling one's temper, and others.

Humans are not all the same even if we believe in equal treatment of all people. We saw that Amerindians were not equal to Europeans in resistance to disease. Their lack of immunity help to decimate them. Inherited genes made the difference. In the broader sense, inherited genes make groups of people different in every conceivable way. Those differences can become large when breeding is confined to within that group. On the other hand, interbreeding of a group with the general population dilutes the trait.

40.3   THE FATE OF HUMANITY

We live in an interesting time. Our innovative capabilities are mind-boggling! The power at our fingertips is both awesome and terrifying! And yet, our primal urges have not been adequately subdued. We no longer kill one another with spears or arrows. Instead, religious zealots fly passenger jetliners into tall buildings, or drones operated from hundreds of miles away eliminate human life with the passionless effort used to swat a fly. Democracy has brought individual freedom to millions, but the issues are too complex for the voters, and special interests spend money to subvert the common good for their own short-term gain. Large parts of the world do not even have freedom of expression or belief. Expressing the wrong opinion in these countries could cost you your life.

We have huge potential, but we have huge problems. The world is becoming smaller due to devices like mobile phones, television, and the Internet. Our innovative ability seems to have raced far ahead of our social ability to deal with the changes. The world population of humans is now so large that catastrophic events await us. Problems like global climate change, massive crop failure, disappearance of the bees, or nuclear weapons falling into the wrong hands could spell disaster. Let's hope that each new generation of humans is evolving skill sets, such that we can live in peace and make the world less dangerous.

40.4   SUMMARY

Mutations occur rarely and when they do occur, they can be either harmful, neutral, or beneficial. Harmful mutations disappear from the species through culling, neutral mutations do nothing, and good mutations help us adapt to changing conditions. This process is independent of stresses acting on the species, but the number of new mutations is directly related to the population size. The human population is now over seven billion people, so the number of good mutations in the germ line must be at record levels. On the other hand, modern civilizations have eliminated most of the dangers that killed most people many generations ago. Yet viruses and drug-resistant diseases can kill millions when they strike, so super-immunity can still give that person an advantage. Improved genes for improved social interaction, learning rate, and handling stress would be beneficial in today's environment. There are changes to the human species that are observable and continue to happen. For example, brain size is getting smaller. Perhaps the brain has rewired itself to be more efficient. The jaw is getting smaller with the result that many people have impacted wisdom teeth. Some people never get those third molars.

PROBLEM SET FOR PART VII

QUESTIONS

Q1.   Why were the Cro-Magnon hunters able to outcompete the Nean-
       derthal hunters?

Q2.   The cave art of France and Spain contains wall paintings of ani-
       mals that have long since gone extinct. Name three of them.

Q3.   Between 40,000 and 11,000 years ago, the Cro-Magnon people in
       Europe seems to have grown in population, lived more comfort-
       ably than bare survival, and had time for art, music, and spiritu-
       alism. This seems odd considering they lived in Ice Age Europe.
       How do you explain it?

Q4.   The Fertile Crescent is an area of arable land surrounded by inhos-
       pitable boundaries. Name the boundaries to the north? south? east?
       and west? What two major rivers run through the Fertile Crescent?

Q5.   What was the Neolithic revolution?

Q6.   Lactose intolerance is the normal condition for adult humans, yet
       many of us are able to digest milk with no problems. How did that
       capability develop?

Q7.   Disease and malnutrition increased markedly during the agricul-
       tural revolution as compared with hunter-gatherers. Why was this
       true?

Q8.   Contemporary with *Homo habilis* and early *Homo erectus* lived
       a different kind of hominin referred to as robust Australopiths or
       Paranthropus. How were they different from gracile Australopiths
       and from *Homo erectus*?

Q9.   Has the human species stopped evolving as a result of modern
       medicine?

BIBLIOGRAPHY

PART I: FOSSILS TELL A STORY

CHAPTER 1: STRATIGRAPHY

- Frank Press, & Raymond Siever, (1999). *Understanding Earth*, New York, W.H. Freeman and Co.

CHAPTER 2: TIME DIVISIONS

- Frank Press, & Raymond Siever, (1999). *Understanding Earth*, New York, W.H. Freeman and Co.

CHAPTER 3: THE K/T EXTINCTIONS AND THE MAMMALIAN SPECIES RADIATION

- Clive Finlayson, (2009). *The Humans Who Went Extinct*, New York, Oxford University Press.
- Charles Officer, & Jake Page, (1996). *The Great Dinosaur Extinction Controversy*. Menlo Park, California, Addison-Wesley Publishing Co, Inc.
- James Laurence Powell, (1998). *Night Comes to the Cretaceous*. New York, W.H. Freeman
- PBS, *The Rise of Mammals*. http://www.pbs.org/wgbh/evolution/library/03/1/l_031_01.html.
- Wikipedia - Evolution of Mammals.
- https://en.wikipedia.org/wiki/Evolution_of_mammals.
- Wikipedia, https://en.wikipedia.org/wiki/Cenozoic.
- The K/T Extinction, http://www.ucmp.berkeley.edu/education/events/cowen1b.html.

CHAPTER 4: PRIMATES AND APES

- Ian Redmond, (2008). *The Primate Family Tree*. Firefly Books Ltd.

PART II: BIPEDAL SPECIES

CHAPTER 5: FOSSILS AND HUMAN EVOLUTION

- Richard Klein, & Blake Edgar, (2002). *The Dawn of Human Culture*. New York, John Wiley & Sons.
- Ian Tattersall, & Jeffrey Schwartz, (2001). *Extinct Humans*. NY, Nevraumont Publishing Co.
- Chris Stringer, (2012). *The Origin of Our Species*. England, Penguin Books.

CHAPTER 6: THE PALEOANTHROPOLOGISTS

- Mary Bowman-Kruhm, (2010). *The Leakeys*. Amherst, N.Y., Prometheus Books.
- Virginia Morrell, (1995). *Ancestral Passions*. New York, N.Y., Touchstone (2015) NOVA video "Dawn of Humanity."

CHAPTER 7: OVERVIEW OF OUR DISTANT ANCESTORS

- Richard Leakey, & Roger Lewin, (1992). *Origins Reconsidered*. New York, N.Y., Anchor Books.
- Charles Lockwood, (2007). *The Human Story*. London, England, The Natural History Museum.
- Ian Tattersall, & Jeffrey Schwartz, (2001). *Extinct Humans*. NY, NY, Nevraumont Publishing Co.

CHAPTER 8: ARDI

- https://en.wikipedia.org/wiki/Ardipithecus.
- http://humanorigins.si.edu/evidence/human-fossils/species/ardipithecus-ramidus.
- Donald Johanson, & Edgar Blake, (2006). *From Lucy to Language*. New York, N.Y., Simon and Shuster.
- Charles Lockwood, (2007). *The Human Story*. London, England, The Natural History Museum.

CHAPTER 9: LUCY AND AUSTRALOPITHECUS AFARENSIS

- Donald Johanson, & Maitland Edey, (1990). *Lucy—The Beginnings of Humankind*. New York, N.Y., Simon and Shuster.

- Donald Johanson, & Edgar Blake, (2006). *From Lucy to Language.* New York, N.Y., Simon and Shuster.
- Dean Falk, (2004). *Brain Dance.* University Press of Florida.
- Charles Lockwood, (2007). *The Human Story.* London, England, The Natural History Museum.
- Ian Tattersall, (1995). *The Fossil Trail.* Oxford University Press.
- Ian Tattersall, & Jeffrey Schwartz, (2001). *Extinct Humans.* NY, NY, Nevraumont Publishing Co.

CHAPTER 10: AUSTRALOPITHECUS SEDIBA

- Charles Lockwood, (2007). *The Human Story.* London, England, The Natural History Museum, pp. 22–30.
- https://en.wikipedia.org/wiki/Australopithecus_sediba.

CHAPTER 11: HOMO NALEDI

- PBS, NOVA National Geographic documentary, *Dawn of Humanity.*
- Bruce Bower, (2016). *Pieces of Homo naledi story continue to puzzle.* Science News April 19.
- Lee Berger, & John Hawks, (2017). *Almost Human.* National Geographic Partners.

CHAPTER 12: HOMO ERECTUS

- Richard Leakey, & Roger Lewin, (1992). *Origins Reconsidered.* New York, N.Y., Anchor Books, G. Philip Rightmire, (1990). *The Evolution of Home erectus.* Cambridge, MA, Cambridge University Press.
- Lars Werdelin, (2013). *King of the Beasts.* Sci. Am.
- Ian Tattersall, (1995). *The Fossil Trail.* Oxford University Press.
- Ian Tattersall, &Jeffrey Schwartz, (2001). *Extinct Humans.* NY, NY, Nevraumont Publishing Co.

CHAPTER 13: NEANDERTHAL MAN

- Paul Jordan, (2001). *Neanderthal.* Great Britain, Sutton Publishing.
- Clive Finlayson, (2009). *The Humans Who Went Extinct.* New York, Oxford University Press.
- Ian Tattersall, (1995). *The Fossil Trail.* Oxford University Press.
- Ian Tattersall, & Jeffrey Schwartz, (2001). *Extinct Humans.* NY, NY, Nevraumont Publishing Co.

- Wm. H. Calvin, (2004). *A Brief History of the Mind*. New York, Oxford University Press.

CHAPTER 14: HOMO SAPIENS

- Brian Fagan, (2010). *Cro-Magnon*. New York, Bloomsbury Press.
- http://freepages.genealogy.rootsweb.ancestry.com/~villandra/McKinstry/GravettianLinks.html.
- Clive Finlayson, (2009). *The Humans Who Went Extinct*. New York, New York, Oxford University Press.

PART III: HOW EVOLUTION WORKS

CHAPTER 15: CHARLES DARWIN

- Charles Darwin, (1859). *On the Origin of the Species*. Penguin Books.

CHAPTER 16: THE MODERN SYNTHESIS

- Ricard Dawkins, (2013). *An Appetite for Wonder*. United States, Harper Collins.
- Matt Ridley, (2000). *Genome*. New York, Harper Perennial, pp. 38–50.

CHAPTER 17: RICHARD DAWKINS, THE DARWIN OF OUR TIMES

- Ricard Dawkins, (2013). *An Appetite for Wonder*. United States, Harper Collins.
- Richard Dawkins, (1976). *The Selfish Gene*. Oxford University Press.

CHAPTER 18: MECHANISMS OF SPECIATION

- Peter, & R. Grant, (1999). *Ecology and Evolution of Darwin's Finches*. Princeton University Press.
- Richard Dawkins, (1987). *The Blind Watchmaker*. W.W. Norton & Co.
- Richard Dawkins, (1997). *Climbing Mount Improbable*. New York, W.W. Norton & Co.
- The Russian Fox Story: http://scienceblogs.com/thoughtfulanimal/2010/06/14/monday-pets-the-russian-fox-st/.

CHAPTER 19: THE RED QUEEN EFFECT

- Matt Ridley, (2003). *The Red Queen*. United States, Harper Perennial.

CHAPTER 20: EVOLUTION OF BIPEDAL APES AND HUMANS

- Donald Johanson, & Edgar Blake, (2006). *From Lucy to Language*. New York, N.Y., Simon and Shuster.
- Donald Johanson, & Edey Maitland, (1990). *Lucy- The Beginnings of Humankind*, New York, N.Y., Simon and Shuster.
- Paul Jordan, (2001). *Neanderthal*. Great Britain, Sutton Publishing.
- Jared Diamond, (2002). *The Rise and Fall of the Third Chimpanzee*. Great Britain, Random House.
- Richard Leakey, & Roger Lewin, (1992). *Origins Reconsidered*. New York, N.Y., Anchor Books.

PART IV: DNA: A POWERFUL NEW TOOL

CHAPTER 21: INTRODUCTION TO DNA TECHNOLOGY

- James D Watson, (1968). *The Double Helix*. New York, N.Y., Touchstone.
- James D. Watson, (2004). *DNA The Secret of Life*. United Kingdom, Arrow Books.

CHAPTER 22: THE RACE TO DISCOVER DNA'S STRUCTURE

- James D. Watson, (1968). *The Double Helix*. New York, N.Y., Touchstone.

CHAPTER 23: DISCOVERING THE SECRET TO LIFE

- James D. Watson, (2004). *DNA The Secret of Life*. United Kingdom, Arrow Books.
- Fredric M. Richards, (1991). *The Protein Folding Problem*. Sci. Am.

CHAPTER 24: MUTATIONS AND JUNK DNA

- Daniel J. Fairbanks, (2007). *Relics of Eden*. Amherst, N.Y., Prometheus Books.
- Matt Ridley, (2000). *Genome*. New York, Harper Perennial, pp. 122 etc.

PART V: DNA APPLIED TO PALEOANTHROPOLOGY

CHAPTER 25: DNA SCIENCE APPLIED TO HUMAN ORIGINS

- Eugene E. Harris, (2015). *Ancestors in Our Genome*. Oxford University Press.
- Matt Ridley, (2000). *Genome*. New York, Harper Perennial.
- Steve Olson, (2002). *Mapping Human History*. New York, N.Y., Houghton Miffin Co.
- Spencer Wells, (2003). *The Journey of Man: A Genetic Odyssey*. Random House Trade Paperback.

CHAPTER 26: TRACING OUR APE HERITAGE

- Eugene E. Harris, (2015). *Ancestors in Our Genome*. Oxford University Press.
- Dean Falk, (2004). *Brain Dance*. University Press of Florida, Chapter 4.

CHAPTER 27: THE AGE AND ORIGIN OF OUR SPECIES

- Richard Dawkins, (1995). *River Out of Eden*. New York, N.Y., Basic Books.
- Eugene E. Harris, (2015). *Ancestors in Our Genome*. Oxford University Press., Chapter 6.
- Steve Olson, (2002). *Mapping Human History*. New York, N.Y., Houghton Miffin Co.
- Spencer Wells, (2003). *The Journey of Man: A Genetic Odyssey*. Random House Trade Paperback.

CHAPTER 28: OUT OF AFRICA

- Spencer Wells, (2003). *The Journey of Man: A Genetic Odyssey*. Random House Trade Paperback, Steve Olson, (2002). *Mapping Human History*. New York, N.Y., Houghton Miffin Co.
- Brian Sykes, (2001). *The Seven Daughters of Eve*. New York, N.Y., W.W. Norton.
- Nicholas Wade, (2006). *Before the Dawn*. Penguin Press.

CHAPTER 29: NEANDERTHAL-HUMAN INTERBREEDING

- Svante Pääbo, (2014). *Neanderthal Man*, New York, N.Y., Basic Books.
- Eugene E. Harris, (2015). *Ancestors in Our Genome*. Oxford University Press.

CHAPTER 30: DENISOVAN-HUMAN INTERBREEDING

- Svante Pääbo, (2014). *Neanderthal Man*. New York, N.Y., Basic Books, Chapters 22 and 23.
- Eugene E. Harris, (2015). *Ancestors in Our Genome*. Oxford University Press.

PART VI: UNIQUELY HUMAN EVOLUTION

CHAPTER 31: BIPEDAL WALKING

- Elaine Morgan, (1994). *The Scars of Evolution*. Oxford University Press.
- Richard Leakey, & Roger Lewin, (1992). *Origins Reconsidered*. New York, N.Y., Anchor Books.
- Dean Falk, (2004). *Brain Dance*. University Press of Florida, Chapter 4.
- Clive Finlayson, (2009). *The Humans Who Went Extinct*. New York, New York, Oxford University Press.

CHAPTER 32: HAIRLESSNESS

- Nina Jablonski, (2013). *Skin, A Natural History*. Berkeley, CA, University of California Press.
- Elaine Morgan, (1994). *The Scars of Evolution*. Oxford University Press.
- Dean Falk, (2004). *Brain Dance*. University Press of Florida.

CHAPTER 33: BIG BRAIN DEVELOPMENT

- Gary Lynch, & Richard Granger, (2009). *Big Brain*. New York, N.Y., Palgrave McMillan.
- Geoffrey Miller, (2000). *The Mating Mind*. New York, N.Y., Anchor Books.

CHAPTER 34: SPEECH AND LANGUAGE

- Matt Ridley, (2000). *Genome*. New York, Harper Perennial, pp. 96–102.
- Eugene E. Harris, (2015). *Ancestors in Our Genome*. Oxford University Press, pp. 105–108.
- David Anthony, Dimitri Telegin, & Dorcas Brown, (1991). *The Origin of Horseback Riding*. Sci. Am.
- Gregory Cochran, & Henry Harpending, (2009). *The 10,000 Year Explosion*. New York, N.Y., Basic Books.

- Richard Dawkins, (1976). *The Selfish Gene*. Oxford University Press, Chapter 11.
- Jared Diamond, (2002). *The Rise and Fall of the Third Chimpanzee*. Great Britain, Random House.

CHAPTER 35: FIRE, COOKING, AND TOOLS

- Richard Wrangham, (2010). *Catching Fire*. New York, N.Y., Basic Books.
- Brian Fagan, (2010). *Cro-Magnon*. New York, Bloomsbury Press.
- Richard Klein, & Blake Edgar, (2002). *The Dawn of Human Culture*. New York, John Wiley and Sons.
- Jared Diamond, (2002). *The Rise and Fall of the Third Chimpanzee*. Great Britain, Random House.

CHAPTER 36: SEX

- Geoffrey Miller, (2000). *The Mating Mind*. New York, N.Y., Anchor Books.
- Jared Diamond, (2002). *The Rise and Fall of the Third Chimpanzee*. Great Britain, Random House.

PART VII: HOMO SAPIENS DOMINATE

CHAPTER 37: GREAT LEAP FORWARD

- Dean Falk, (2004). *Brain Dance*. University Press of Florida, Chapter 8.
- Ian Tattersall, (1998). *Becoming Human*. New York, N.Y, Harcourt Brace and Co.
- Richard Klein, & Blake Edgar, (2002). *The Dawn of Human Culture*. New York, John Wiley & Sons.
- Spencer Wells, (2003). *The Journey of Man: A Genetic Odyssey*. Random House Trade Paperback, Gregory Cochran, & Henry Harpending, (2009). *The 10,000 Year Explosion*. New York, N.Y., Basic Books.

CHAPTER 38: AGRICULTURE AND CIVILIZATION

- Gregory Cochran, & Harpending Henry, (2009). *The 10,000 Year Explosion*. New York, N.Y., Basic Books.
- Jared Diamond, (2005). *Guns, Germs, and Steel*. New York, N.Y., W.W. Norton and Co.

CHAPTER 39: THE JOURNEY FROM APE TO MAN

- Gregory Cochran, & Henry Harpending, (2009). *The 10,000 Year Explosion.* New York, N.Y., Basic Books.
- Jared Diamond, (2005). *Guns, Germs, and Steel.* New York, N.Y., W.W. Norton and Co.
- Richard Klein, & Blake Edgar, (2002). *The Dawn of Human Culture.* New York, John Wiley and Sons.

CHAPTER 40: WHAT THE FUTURE HOLDS

- Gregory Cochran, & Henry Harpending, (2009). *The 10,000 Year Explosion.* New York, N.Y., Basic Books.
- John R. Skoyles, & Dorian Sagan, (2002). *Up from Dragons.* McGraw-Hill.

INDEX